D1327052

WHITAKER'S ALMANACK
LITTLE BOOK OF
ASTRONOMY

MIKE FLYNN

A & C BLACK · LONDON

Whitaker's is a registered trade mark, Registered Trade Mark Nos. (UK) 1322125/09; 13422126/16 and 13422127/41; (EU) 19960401/09/41, licensed for use by A&C Black (Publishers) Ltd.

A&C Black Publishers Ltd
38 Soho Square
London W1D 3HB
Tel: 020-7758 0200
Fax: 020-7758 0222
Web: www.whitakersalmanack.com

Copyright © 2006 Elwin Street Limited

Conceived and produced by
Elwin Steet Limited
79 St John Street
London EC1M 4NR
www.elwinstreet.com

A CIP catalogue record for this book is available from the British Library.

ISBN-10: 0-7136-8237-6
ISBN-13: 978-0-7136-8237-3

Cover images © PhotoDisc
Printed and bound at Star Standard Industries in Singapore.

Contents

89   THE STUFF OF THE UNIVERSE

99   LOOKING AT GALAXIES

115   SPACE EXPLORATION

ORIGINS

The size of the Universe

What do you see when you look up into the night sky? Do you marvel at the beauty of this fiery celestial ballet or simply wonder at the sheer scale of the heavens? Whatever thoughts or emotions cross your mind, there really is no getting away from the fact that the Universe is an awfully big place.

In truth, our brains are not really equipped to comprehend the sheer size of the Universe. Words like 'enormous', 'gigantic' or even 'awesome' simply don't come close to conveying a true sense of scale, so let's start small and work our way up.

Imagine that our entire Solar System – basically the Sun, the planets and a few leftover bits and pieces – is laid out on the Centre Court at Wimbledon in London, England, with Pluto at one end and the Sun at the other. Working at this truly tiny scale, how far away do you imagine the next nearest star would lie?

If you said Johannesburg, South Africa, you were probably reading ahead, but you'd also be right – and that's only our nearest neighbour. Even travelling at the speed of light, which would allow you to circle the Earth seven times in just one second, it would still take nearly four years to get there. That's a long way to go and also the reason why we tend to measure distances in space in terms of the speed of light.

Travelling through the vacuum of space, light moves at around 300,000 km (186,000 miles) per second, which seems very impressive. But even at this speed it would take 100,000 years to cross the Milky Way Galaxy, which is home to our Sun and all of the stars that you can see in the night sky. The journey to the next nearest galaxy would take an incredible 179,000 years, even at the speed of light. But what about the big one?

According to current estimates, cruising at the speed of light – which, let us remind ourselves, is an incredible 300,000 km (186,000 miles) per second – it would take something like 13.7 billion years to cross the known Universe.

Big, isn't it?

A brief history of the Universe

We know for certain that the Earth and the other planets in our Solar System orbit a raging inferno which we call the Sun. The Sun is a fairly ordinary star, one of about five hundred billion or so stars held in orbit in a loose spiral around what is almost certainly a truly massive black hole. Astronomers have named this spiral-shaped collection of planets, stars and assorted leftovers the Milky Way Galaxy.

Our Sun is tucked away in one of the arms of this spiral galaxy – the cosmic equivalent of living in Alaska. The spiral rotates at the rate of once every 230 million years or so, slow enough that these stars appear to be fixed in the night sky and can be used for navigational purposes. But it has taken us quite some time to reach this understanding of our place in the Universe.

The Greeks and Romans

The first recorded scientific theory on the nature of the Universe was put forward by the ancient Greeks, such as Thales and Anaximander. During the sixth century BCE they dropped the notion of a supernatural origin (i.e., 'God made it') and proposed that the Universe had been created by natural means. The Pythagoreans agreed, creating a mathematical model of the Universe that was adopted by the Romans.

The Roman poet Lucretius, in *De Rerum Natura* (On the Nature of Things), wrote of an infinite Universe in which the interaction of things at the atomic level created a constant state of flux. This was a truly remarkable effort considering just how similar it is to modern thinking on the nature of the Universe.

Although Aristotle and Plato also made contributions, the next major school of thought was led by Ptolemy. He described a finite Universe ruled by mathematics and God in which the Sun, planets and other stars were attached to concentric spheres centred on the Earth. He could not have been more wrong, but unfortunately his views were adopted by all-powerful medieval theologians.

ABOVE Frontispiece from Ptolemy's *Almagest* of 1496 that described the Ptolemic system of astronomy and geography.

ABOVE Frontispiece of Galileo's *Cosmic System* of 1663, featuring Aristotle, Ptolemy and Copernicus on the illustration shown here.

Heliocentric Universe

That was pretty much it until the sixteenth century when Nicolaus Copernicus revived the Greek idea that the Sun, and not the Earth, was at the centre of the Universe. Copernicus' theory was supported by the systematic and highly accurate observations of astronomer Tycho Brahe (pronounced 'Bra-hey') and adopted by Galileo Galilei. Galileo promoted this theory as his own, helping to spread it among European thinkers. He also came close to losing his life at the hands of the Church, who saw his ideas as heretical. Eventually he recanted but ended his days under house arrest.

Newton and the mechanists

Gradually, and inevitably, the truth became obvious and by the time Isaac Newton had provided a mathematical, mechanistic basis for our understanding of the Universe reason had finally triumphed over superstition. Newton's influence cannot be overestimated. Here was a man who, to his contemporaries, seemed to be able to read the mind of God, explaining His mysteries using mathematical proofs. Quite rightly, Newton's theories on the nature of the Universe remained dominant until the beginning of the twentieth century, when Albert Einstein introduced his theories on relativity.

The expanding Universe

Even before Einstein, however, some astronomers were beginning to notice that something was amiss. By the early twentieth century, improvements in the size and quality of telescopes allowed astronomers to see further into space than ever before and this improved resolution was turning up more than a few surprises.

The notion that the Universe was expanding had been around for a while. But in 1929 American astronomer Edwin Hubble was able to show that not only were all the galaxies in the Universe (including our own) moving at tremendous speeds, the ones that were furthest away from us were moving away the fastest. Suddenly the Universe was a very, very big place.

Big bangs and steady states

Having established a rough idea of the size of the Universe, astronomers began once again to turn their attention to perhaps the biggest question of all: how did we get here? First in line to answer this question was, not surprisingly, a man of God.

In 1931, Belgian priest and scientist Georges–Henri Lemaître proposed the theory that the expansion of the Universe was a consequence of the spontaneous disintegration (i.e., explosion) of what he called the 'primal atom', a single entity containing all of the matter and energy in the Universe. In short, he claimed that it all began with a great big bang.

Although now the widely accepted view of the origin of the Universe, this theory did not find universal acceptance when first proposed. The English astronomer Fred Hoyle even mocked the theory and was clearly unwilling to believe that the Universe began with a 'big bang'. (Ironically, in using this term he managed to give the theory its popular name.) Hoyle, and others before him, believed that the Universe was in a steady state.

By this he meant that the Universe, although constantly expanding, had no beginning, will have no end, and when viewed from any point in time or space would maintain an average density because new matter was being created continuously to fill the growing spaces.

Unfortunately for Hoyle and his supporters, compelling evidence supporting the Big Bang theory appeared in 1965. Arno Penzias and Robert Wilson, a couple of radio astronomers from the

FACT
A millionth of a trillionth of a trillionth of a second after the Big Bang the temperature of the Universe was a whopping 100,000 billion billion billion°C.

FACT

Hydrogen, which may soon replace petroleum as fuel, came into existence around 100 seconds after the Big Bang.

US, struck academic gold when they accidentally stumbled across microwave background radiation – essentially an echo of the original Big Bang explosion – while trying to eliminate radio interference.

The Steady State Theory was consigned to the dustbin of history at just about the point where Penzias and Wilson picked up their Nobel Prize, an astonishing result for them considering that they thought the interference they'd detected was caused by pigeon droppings building up in their radio telescope.

Other universes

Of course, the above models are not the only theories on the origins and nature of the Universe. Leaving aside religious beliefs, there have been several intriguing ideas put forward in recent years. Among the most interesting is the philosophical idea of the Anthropomorphic Universe. Put simply, this theory states that the Universe is the way it is because the act of observing it makes it that way. In short, we define reality.

Another idea, and one that is increasingly gaining acceptance among mathematicians, particle physicists and astronomers is the notion of a Multi-dimensional Universe. There is certainly mathematical evidence to support the possibility that there are at least 11 dimensions (though some claim up to 26), the 4 that we can perceive – length, width, height and time – plus another 7 which are beyond our senses (and most people's comprehension). The universe measured along these additional dimensions is subatomic in size. As ever, only time and a few very fine minds will tell if either of these theories has any real validity.

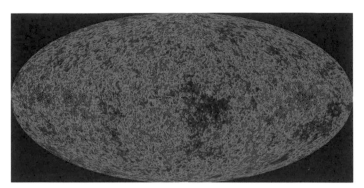

ABOVE A full-sky map of the oldest light in the Universe. Colours indicate 'warmer' (red) and 'cooler' (blue) spots.

How it may end

So, having examined how we may have got here, it remains only to decide on the final act. Will it all end with a bang or a whimper? The answer depends on just how much 'stuff' there is in the Universe.

If there is enough matter then ultimately gravity will take hold and the current expansion will cease, giving way to a gradual contraction that will result in what has been dubbed the 'Big Crunch'. The fact that this may in turn trigger another big bang has been keeping philosophers, astronomers and sci-fi fans awake at night. It's an exciting idea, especially when one considers that we have no reason to believe that this will be the very first time that this has happened.

It is, however, thought more likely that there is not enough matter in the Universe for gravity to pull it all back together, and so the Universe will simply continue to expand indefinitely. Ultimately, all of the energy in existence will dissipate until the Universe will, effectively, be dead.

But don't worry. That's not going to happen anytime soon. By then our solar system and everything in it, from Shakespeare to *The Simpsons*, will be long gone.

STUDYING
SPACE

Optical telescopes

Despite being a truly remarkable organ, the human eye makes a poor telescope. Its small opening (the aperture) doesn't let in a great deal of light and the degree of magnification is fixed. The telescope was designed to overcome these two basic design limitations and allow us to look beyond the scope of our own eyes.

Optical telescopes – the type most people think of when they imagine a telescope – possess a much larger aperture than the human eye. This means that they can be used to collect much more of the light coming from distant objects, which greatly improves resolution and clarity.

Magnification is less of an issue with optical telescopes that are trained on the night sky. The Sun aside, even the nearest stars are so far away that no amount of magnification is really going to make them appear as anything other than points of light against the blackness of space.

How optical telescopes work

Optical telescopes can be divided into two different types: refracting and reflecting.

The refracting telescope

The refracting telescope was the earliest design and is usually formed using two lenses. The distance between the two lenses, which are commonly placed near either end of a tube, can be adjusted to vary the resolution and magnification required. Any light passing through the forward lenses is refracted (bent) before being focused on the eyepiece lens.

The reflecting telescope

The reflecting telescope produces an image with the aid of a large concave mirror. Light is collected by the mirror and focused on a smaller secondary mirror before passing through to the eyepiece lens. The reflecting telescope is also referred to as the Newtonian, after the English astronomer Isaac Newton who first used this design to build a telescope around 1670.

Electromagnetic spectrum

As can be seen from the illustration below, visible light represents just a tiny portion of the electromagnetic spectrum. Telescopes that can 'see' across the whole range of the spectrum were developed so that we can gain insights into the workings of much more of the Universe than was previously possible.

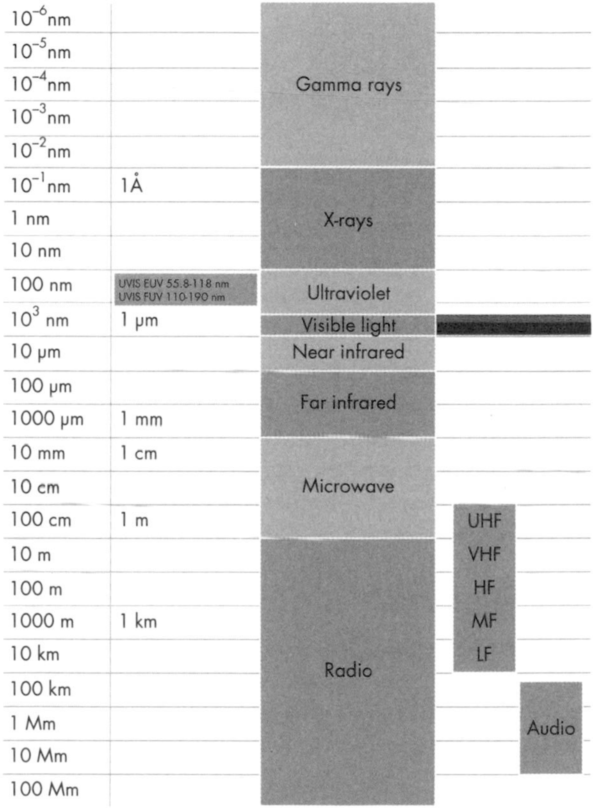

10^{-6} nm		Gamma rays	
10^{-5} nm			
10^{-4} nm			
10^{-3} nm			
10^{-2} nm			
10^{-1} nm	1 Å	X-rays	
1 nm			
10 nm			
100 nm	UVIS EUV 55.8-118 nm / UVIS FUV 110-190 nm	Ultraviolet	
10^3 nm	1 µm	Visible light	
10 µm		Near infrared	
100 µm		Far infrared	
1000 µm	1 mm		
10 mm	1 cm		
10 cm		Microwave	
100 cm	1 m		UHF
10 m			VHF
100 m			HF
1000 m	1 km		MF
10 km		Radio	LF
100 km			
1 Mm			Audio
10 Mm			
100 Mm			

nm = nanometre; Å = angstrom; µm = micrometre; mm = millimetre;
cm = centimetre; m = metre; km = kilometre; Mm = Megametre

ABOVE Electromagnetic spectrum with highlighted narrow band of visible light.

LEFT Scale model of Great Rosse reflecting telescope of 1842, used to determine the nature of nebulae.

Non-optical telescopes

Non-optical telescopes allow observations to be made beyond the narrow band of visible light that we perceive. Astronomers use a range of telescopes which are tuned to different parts of the electromagnetic spectrum. These allow them to 'look' directly at the Sun, 'see' the natural radio signals from distant stars or, by using X-ray and gamma ray telescopes, construct images of distant black holes.

Astronomical telescopes

Astronomers use a range of telescopes when studying the Universe. Each is tuned to a particular part or parts of the electromagnetic spectrum and while many are Earth-based, the trend is toward placing ever more telescopes in space. These are usually to be found in orbit around the Earth.

Measuring distances in space

How do astronomers measure distances in space? They can't very well go out there with a very long tape measure to check lengths and widths and a pedometer is never going to suffice. Instead, they have come up with many clever and cunning ways of measuring these vast expanses.

One of the earliest methods of measuring distances in space effectively turns the Universe into a vast trigonometry problem.

Parallax shift

When seen from different points on the Earth's orbit around the Sun, nearby stars can appear to shift against the backdrop of the night sky. This apparent movement is called the parallax shift. By using simple trigonometry it is possible to calculate fairly accurately the distance to these relatively nearby stars.

Cepheid variables

Astronomers measure the distances to more distant stars by comparing a star's true brightness (or Absolute Magnitude) with its brightness as seen from Earth (its Apparent Magnitude). Cepheid stars are very bright objects that appear to pulsate. The time taken for the star to pulsate gives its true brightness. By measuring the difference between this and its apparent brightness it is possible to calculate the distance to that star. This is the most common method of determining distances to faraway galaxies.

The Hubble Space Telescope

Launched on 25 April, 1990, the Hubble Space Telescope opened up the unseen Universe to human eyes, allowing us to view objects seven times more distant than previously possible using the most powerful telescopes here on Earth.

Essentially a giant reflecting telescope in orbit 966 km (600 miles) above our planet, Hubble sits well above the distorting effects of the atmosphere and offers resolutions 10 times sharper than those obtained with ground-based telescopes.

FACT

The distorting effects of our planet's atmosphere are such that even an ordinary hand-held telescope would be more effective when used in space than the most powerful optical telescope currently based here on Earth.

In addition to its 2.4-m (94-inch) primary mirror, Hubble is also fitted with cameras that are sensitive to visible light as well as infrared and ultraviolet radiation. Images are collected on these cameras using a charge-coupled device before they are transmitted down to Earth for analysis.

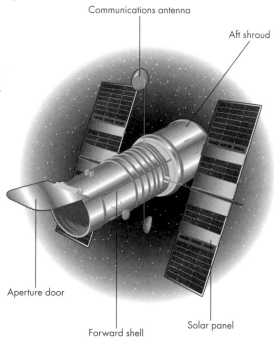

Communications antenna

Aft shroud

Aperture door

Forward shell

Solar panel

ABOVE The body of the Hubble Space Telescope.

The Universe through Hubble's eyes

The launch of the Hubble Space Telescope was a truly momentous occasion in the history of astronomy. For the first time ever we were able to see with our own eyes the true majesty of the Universe. From star birth in the Eagle Nebula to primitive early galaxies at the edge of creation, at a single stroke Hubble removed the shades from our eyes and changed our view of the Universe forever.

LEFT Hubble Space Telescope image of a dying star.

Why is it called the 'Hubble' Space Telescope?

The Hubble Space Telescope is named after Edwin Hubble (1889-1953). He was a Missouri-born lawyer-turned-astronomer who is most famous for revealing something of the true scale of the Universe. His work on distant nebulae (vast interstellar clouds of gas) revealed that many of these apparent clouds are in fact galaxies lying at incredible distances from us. Hubble's observations established beyond doubt that these fuzzy 'nebulae' were not part of our galaxy, as had been thought, but were galaxies themselves, outside the Milky Way. He is also noted for having devised a classification system for galaxies, grouping them according to their content, distance, shape, size and brightness.

FACT

The Hubble Space Telescope failed to impress on its first outing before it was realised that its reflecting mirror, the key component in the telescope, was misshaped. Astronauts had to fit the telescope with what is in effect a giant contact lens in order to correct the fault.

Worlds beyond – planets of other suns

Having established where we are and how long we've been here – if not yet why we are here – scientists inevitably turned their attention to perhaps the most important question ever to face the human race: are we alone?

It seems inconceivable to the human mind that, given the vast scale of the Universe, we should be the only intelligent inhabitants. But, to use the words of a well-known particle physicist and sceptic Enrico Fermi, where is everybody?

Back in the 1960s, the search began for Earth-like planets in orbit around distant stars. At the time, telescopes were not powerful enough to be able to pick out an Earth-sized planet anywhere outside of our Solar System – the light coming from the star it would

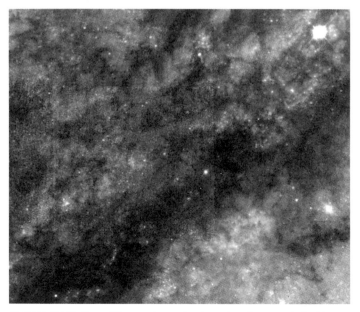

ABOVE A Hubble Space Telescope image showing galaxy Centaurus A, Earth's nearest active galaxy.

be orbiting would also make it all but invisible. This situation has barely improved since then. So, how do we find a distant planet if we have no chance of seeing it?

Star wobble

Everything with mass can exert a gravitational influence on anything else with mass. The Earth, for example, is kept in orbit by the gravitational pull of the Sun, which in turn experiences a corresponding 'tug' from the Earth as the Earth moves along its orbit. Although ordinarily we cannot perceive this, it is possible to tell the degree to which the Sun is being influenced by the gravitational pull of the Earth by measuring the amount that the Sun 'wobbles' on its axis.

Each orbit of the Earth produces a regular wobble which, even if our home planet could not be seen, would indicate to a distant observer that there is a planet here. This same technique can be applied across space to any observable star, and since 1995 has been turning up more than a few surprises.

The easiest planets to spot are the large (i.e., Jupiter-sized) bodies, which cause a pronounced and relatively easily detected wobble on their parent star. So far around 200 of these worlds have been detected. Improvements in this technique have more recently been producing evidence of Earth-sized worlds in orbit around distant stars. Alas, so far we have failed to detect any form of life on these or any other worlds.

FACT
In an attempt to communicate with potential alien civilisations, a radio signal was sent from the Arecibo telescope in Puerto Rico to a star cluster that lies 25,000 light years away. The earliest we are likely to receive a reply (if at all) will be in 50,000 years' time.

SETI

SETI – the Search for Extraterrestrial Intelligence – is the name adopted by a group of scientists who are dedicated to searching for evidence of intelligent life elsewhere in our galaxy. They use radio telescopes to collect signals from space which are then analysed for potential signs of life. This might, for example, be a repeating pattern with no known natural cause.

With the exception of one signal detected back in the 1970s (and subsequently named the 'Wow!' signal) there has so far been no evidence of intelligent life detected anywhere in the Universe other than on our planet, where its existence or otherwise is still a topic of much debate.

The Drake Equation

Frank Drake, a leading figure in the search for extraterrestrial intelligence, devised an equation (also known as the Green Bank equation) to determine the likelihood of discovering intelligent life elsewhere in our galaxy.

The equation reads:

$$N = R \times f_p \times n_e \times f_1 \times f_i \times f_c \times L$$

At first sight this equations looks daunting but the equation breaks down as follows:

N is the number of intelligent civilisations that could contact us. It is equal to the number of stars in our galaxy (R) multiplied by those with planets that could sustain life (f_p), further multiplied by those that are Earth-like (n_e), the probability that life will begin on such a planet (f_1), the chance that this life will develop technology (f_i), and have the desire to communicate with us (f_c). The final figure, L, represents the average life expectancy of such a civilisation.

Using this equation, Drake predicted that there might be at least 40 such planets currently in our galaxy.

THE
SOLAR
SYSTEM

A solar system

A solar system is a self-contained unit consisting of at least one star which is orbited by one or more planets. There are, of course, variations on this model to be found in the Milky Way Galaxy but this definition covers the basics.

Our Solar System consists of the Sun (our star) plus at least nine planets, their various moons and assorted debris, much of which can be found in the 10,000 or so asteroids that live in the Asteroid Belt. We are only now discovering that our solar system may not be unique, although so far it remains the only place in the Universe that is home to life as we know it.

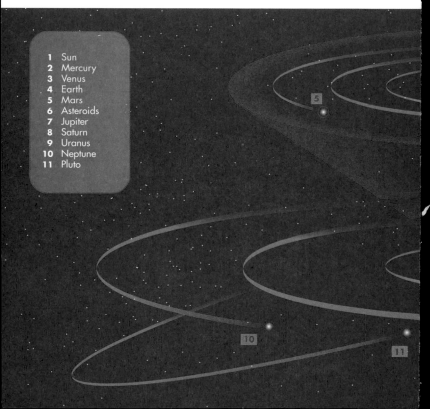

1   Sun
2   Mercury
3   Venus
4   Earth
5   Mars
6   Asteroids
7   Jupiter
8   Saturn
9   Uranus
10 Neptune
11 Pluto

There is some debate as to the number of real planets in our solar system. The Japanese, for example, dismiss Pluto as little more than an oversized lump of rock and ice. More recently, several American astronomers have asserted that up to another dozen planets may lie in a halo-like formation on the outer edge of our solar system, although there is no conclusive proof of this theory . . . yet.

Almost our entire solar system is taken up by the Sun, which accounts for 99.86 per cent of all the mass in the system. Jupiter, the next nearest-sized body in terms of mass, accounts for nearly all of the other matter in the system with the remainder forming the rest of the planets, moons, asteroids, meteors and comets.

Birth of a solar system

Our Solar System was born around five billion years ago in the fierce fires of a nuclear reaction. Over tens of millions of years a vast cloud of gas and dust had gradually built up, particle crashing against particle, clump against clump, each drawn by gravity to the centre of a cloud. Over time the cloud started to spin, faster and faster, until the centre began to bulge and our embryonic solar system took on the form of a spinning top.

All the while the central region, the core, was becoming increasingly dense, so dense that temperatures and pressures rose to a critical point and the core began to undergo nuclear reactions. Truly, a star was born.

The vast spinning disc of gas and dust surrounding the galaxy's latest star continued to rotate and over millions of years the same process that had given birth to our sun began to be repeated across space. Each of the inner planets, Mercury, Venus, Earth and Mars, began life as a few grains of dust, building up over time and combining with other dust particles to form clumps of matter. These clumps of matter inevitably attracted other clumps of matter and it was in this haphazard and random fashion that the planets were formed.

Beyond Mars (the fourth rock from the Sun) lies the Asteroid Belt, a collection of rocky and metallic objects ranging in size from a single dust particle up to huge boulders hundreds of kilometres in

FACT

The hottest surface temperature in the Solar System (other than the surface of the Sun) is to be found on Venus, which has an average temperature of 480°C (896°F). The coldest-recorded surface temperature in the Solar System is that of Neptune's moon Triton, which is a somewhat chilly -235°C (-391°F).

JUPITER

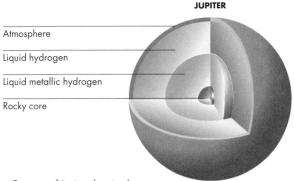

Atmosphere

Liquid hydrogen

Liquid metallic hydrogen

Rocky core

ABOVE Cutaway of Jupiter showing layers of gas, liquid gas and a rocky core.

diameter. Some see the Asteroid Belt as a lost opportunity, a 'wanna-be' planet that never had a hope of survival. Certainly, with the mighty Jupiter as a near neighbour there was little chance that the belt would ever yield a planet.

It's been claimed, and with some justification, that Jupiter is a failed Sun. It was clearly a star in the making with a mass two and a half times that of all of the other planets in the Solar System combined. But there was insufficient matter in the Solar System to generate and support two stars and consequently Jupiter's light remained hidden.

As the first and largest of the planets known as the gas giants, Jupiter's rocky core is clothed in an inconceivably dense cloud of hydrogen and helium. Pressures at the centre of the planet are such that the very core is believed to be sheathed in metallic hydrogen, an incredibly exotic material that can only exist at pressures we cannot hope to reproduce here on Earth.

Saturn, with its famous rings – which are thought to be the debris from a failed or destroyed moon – lies beyond Jupiter and was believed to be at the limit of our Solar System until the discovery of Uranus in 1781. Our horizons shifted further with the discovery of Neptune in 1846 and reached their current limit when shivering little Pluto was spotted in 1930.

Orbits

Strictly speaking, an orbit is a curved path, usually elliptical in shape and described by one object in motion around another under the influence of gravity. When we think of orbits we tend to imagine them on a grand scale – the Earth around the Sun, for example, or the Solar System around the core of the Milky Way Galaxy. But they also occur at the subatomic level, an example of which being the motion of an electron around the nucleus of an atom.

For many years it was taught that the Sun and the planets were in orbit around the Earth. This certainly seemed true to the casual observer but was a fundamentally flawed observation. First among modern thinkers to point this out was a fifteenth-century Polish medic called Mikolaj Kopernik.

Better known by the Latinised version of his name, Nicolaus Copernicus proposed that the Earth rotated daily on its own axis and moved around the Sun in a year-long orbit. He published his ideas in a famous work called *De Revolutionibus Orbium Coelestium* (On the Revolutions of the Celestial Spheres), a ground-breaking publication that unfortunately contained a number of errors.

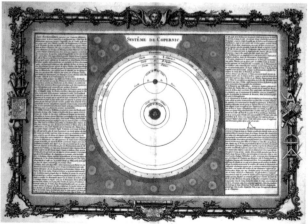

ABOVE Copernicus' model of the Solar System of 1761.

FACT

Copernicus' *De Revolutionibus Orbium Coelestium* (On the Revolutions of the Celestial Spheres) was published in 1543, as he lay on his deathbed. It was banned by the Roman Catholic Church and remained so until 1835.

Among these was Copernicus' description of the circular orbits of celestial bodies. In short, the maths simply didn't work. It took the intervention of a German-born astronomer, Johannes Kepler, to put things right.

Like Copernicus, Kepler was a strong advocate of the heliocentric (sun-centred) Universe but believed that the planets orbited the Sun in elliptical rather than circular orbits. In 1609 he published *Astronomia Nova* (The New Astronomy) in which he showed that the planets move in gentle ellipses around the Sun and that a (very long) line drawn between the Sun and an orbiting planet would sweep out equal areas of the ellipse in equal time. These two assertions are now known as Kepler's First and Second Laws.

Kepler's Third Law, published in 1619 in *Harmonice Mundi* (Harmony of the World), established that there is a distinct relationship between the distance of a planet from the Sun and the time taken for it to travel along its orbit. In essence this law states that a planet's journey along its elliptical path slows the further it travels away from the Sun, only to speed up again as it draws closer in its orbit.

ABOVE Kepler's model of the Universe, drawn up in 1619.

Types of orbiting bodies

Orbiting bodies range from the grand to the subatomic. For astronomers, the main examples are planets, moons, asteroids, comets and meteors.

Planets

Within our Solar System, the largest bodies in orbit around the Sun are the planets. These range in size from the mighty Jupiter, which has a diameter of 141,974 km (88,734 miles) down to tiny Pluto, which at 2,390 km (1,493 miles) could be accommodated in the middle of Europe and as such hardly qualifies as a planet at all.

Moons

Even Pluto has a moon. In fact, with the exceptions of Mercury and Venus, all of the planets have at least one moon. Saturn, in addition to

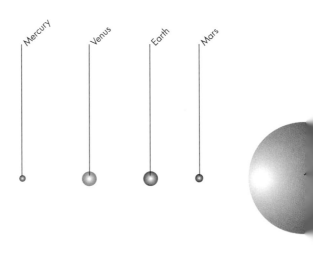

ABOVE A diagram showing the relative sizes of the planets, all of which are dwarfed by the Sun.

its magnificent rings, has the highest number of moons, with 34, and 13 newly discovered satellites confirmed so far. While the planets orbit the Sun, a moon orbits the parent planet. The best example of this is Jupiter, which, with its four principal moons – Io, Callisto, Europa and Ganymede – is essentially a solar system that has failed to fire up.

ABOVE Jupiter's four major moons, Io, Callisto, Europa and Ganymede.

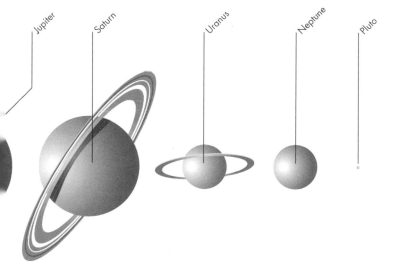

Jupiter   Saturn   Uranus   Neptune   Pluto

Asteroids

The next biggest bodies in orbit around the Sun are the asteroids. On the whole, these tend to be quite small. However, some are up to 1,000 km (620 miles) in diameter. Most asteroids can be located in a belt that lies between the orbits of Mars and Jupiter.

ABOVE The asteroid Eros.

Comets

Comets, which are little more than huge dirty snowballs, tend to live in the dark icy hell that lies beyond the orbit of Pluto. Occasionally the gravity of a passing star will knock one out of position and send it on a very long orbit around the Sun. On Earth we are from time to time blessed with the sight of a celestial snowball melting slightly as it passes near the Sun, leaving behind a characteristic cometary tail.

Meteors

Meteors really are the Solar System's leftovers. These lumps of stone and iron, drifting freely across space, graduate in status to 'meteorites' on contact with our atmosphere. Most burn up on entry, though a few small meteors regularly land on our planet's surface. The largest, however, have the capacity to destroy all life on Earth and have in the past been blamed for key events in our planet's history, including the extinction of the dinosaurs.

FACT

The Torino Scale rates the chances of an object – such as an asteroid, meteor or comet – colliding with Earth across a range of 0 to 10, with 0 indicating no chance of a collision and 10 indicating no chance of humanity surviving such a collision.

What are stars made of?

A star, of which our Sun is a far from remarkable example, is in essence one huge nuclear reactor made of gas. Of course, when we on Earth think of gas we imagine a barely tangible cloud of matter that tends to drift wherever the wind takes it. Gas, as it occurs in stars, could not be more different.

The Sun is composed mostly of liquid hydrogen, which accounts for about 70 per cent of its mass. The rest is helium, although traces of other, heavier elements can also be detected. Like all other stars, the Sun's energy and, therefore, most of the energy in our Solar System, comes from the thermonuclear fusion of hydrogen to form helium. This occurs at the core of the star and the resulting heat is then carried by convection to the surface from where it is radiated out into space. This sounds like a straightforward process but it actually takes around one to two million years for the heat at the core of our Sun to reach the outer layers of its atmosphere.

Although we tend to think of stars as single entities, very many exist in combinations of two or even three, each caught by the others' gravitational influence and trapped in a dance to the death lasting millions or billions of years. Inevitably, stars eventually burn themselves out and just like the stars of the entertainment world, those that burn brightest tend to burn fastest.

What are planets made of?

Planets are large bodies with dense rocky cores. They orbit a star or stars at regular intervals and on elliptical paths. The rocky core may be surrounded by a less dense mantle (in the case of the Earth) or might have a thick, gaseous covering that is held in a liquid state under extreme pressure. The latter typically describes all of the gas giants – Jupiter, Saturn, Uranus and Neptune – in our Solar System.

Planets generally have atmospheres (although one can only infer an atmosphere on the surface of Pluto). These range from the near negligible in the case of Mercury through to the car-crushing dense greenhouse-gas atmosphere of Venus.

FACT

The Asteroid Belt has all the right ingredients for a planet but it would be a very small one. By comparison, our own Moon possesses more mass than the whole of the Asteroid Belt combined.

We call the time taken for the Earth to orbit the Sun 'one year'. But a year on the other planets in our solar system is longer or shorter depending on the planet's location. Mercury, which lies closest to the Sun, takes just 88 days to complete a single orbit while Pluto, the furthest planet from the Sun, takes well over 248 of our years to travel just once around its orbit.

What are moons made of?

A moon is a relatively large body which orbits a planet in much the same way that a planet orbits a star, but on a far smaller scale. Our own Moon is a rocky sphere that we can see when it reflects sunlight down on to us. Despite having a diameter about a quarter that of the Earth, like most moons it lacks sufficient mass to hold on to an atmosphere of its own – in essence the Moon is one giant piece of pumice stone and lacks the surface gravity needed to hold on to gas particles.

Ganymede is the largest of Jupiter's moons. If Ganymede orbited the Sun, rather than Jupiter, it would be classified as a planet. Like our Moon, it has no known atmosphere, but recently the Hubble Telescope detected ozone on its surface, so it is likely to have a thin oxygen atmosphere.

LEFT Ganymede.

The Sun

From its less than promising origins as a swirling cloud of gas and dust our Sun has grown into the medium–sized mature star that we know today. Formed as a truly ancient gas cloud collapsed under the influence of its own gravity, thus igniting the fusion process, the Sun has been blazing away for around five billion years and should be good for at least as long again.

The structure of the Sun

The core of the Sun has a diameter of approximately 450,000 km (280,000 miles) and a density roughly 160 times that of water. It is here that the thermonuclear fusion of hydrogen (to form helium) releases the energy that we have all come to rely on. As this occurs, 0.7 per cent of the mass being fused is converted into energy, producing temperatures at the core in excess of 15 million°C (27 million°F).

Above the dense inner core is a region known as the radiative layer. Forming 70 per cent of the radius of the Sun, this is the region where the heat from the core begins its one to two million-year journey upward to the star's atmosphere.

The radiative layer gives way to the convective layer. This is a 200,000 km (125,000 mile) thick region where heat from the core is carried upwards by streams of gas. As it travels through this layer the temperature drops from 2 million°C (3.6 million°F) to emerge on the surface of the star at a cooler 5,700°C (10,300°F).

FACT
The Sun burns up hydrogen at a rate of around 700 million metric tons per second, yet a single gram of hydrogen fusing to form helium produces as much energy as 600 million electric heaters might radiate in a single second.

FACT

At around 1.4 million km (875,000 miles) in diameter, the Sun is 109 times wider than the Earth and could house over a million Earth-sized planets within its boundaries.

The first layer of the Sun's atmosphere is called the photosphere. This is the bright surface of the star that we can see here on Earth. Perversely, having dropped, temperatures begin to pick up once again in this layer. They will eventually peak at temperatures of about 7,500°C (13,500°F).

Visible as a distinctive pinkish-coloured layer during a total eclipse, the chromosphere lies above the photosphere and is home to temperatures of up to 20,000°C (36,000°F). The chromosphere is shrouded in a corona of hot gases. This corona is up to several million kilometres thick and temperatures can climb as high as 2 million°C (3.6 million°F) before the light and heat that began their journey 2 million years earlier are cast out into space.

Solar flares

Angry, violent flashes of magnetic energy tear through the Sun's chromosphere flinging charged particles out into space. Usually lasting no more than 20 minutes or so, these cosmic tantrums are commonly called solar flares. The amount of energy released is the equivalent of millions of 100-megaton hydrogen bombs exploding at the same time.

Sunspots

Super strong magnetic fields on the surface of the Sun produce groups of dark patches known as sunspots. They seem dark because they are cooler than surrounding areas. Appearing mostly around the equator, sunspots can last for as long as six months.

Prominences

Breathtaking clouds of raging gas roar across the Sun's upper chromosphere forming arcs of fire known as prominences. Sometimes, if the fields become unstable, the prominences will erupt and rise off the Sun in just a few hours.

Solar cycle

Magnetic fields act to slow the flow of heat to the surface of the Sun, creating solar flares, sunspots and prominences in an 11-year tide known as the solar cycle.

BELOW The dark patches shown are the Sun's sunspots.

BELOW This spectacular image shows the Sun's fiery prominences.

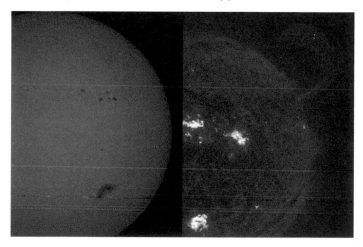

FACT
When you look at the Sun you are effectively looking back in time as light from the Sun takes 8.3 minutes to reach us here on Earth.

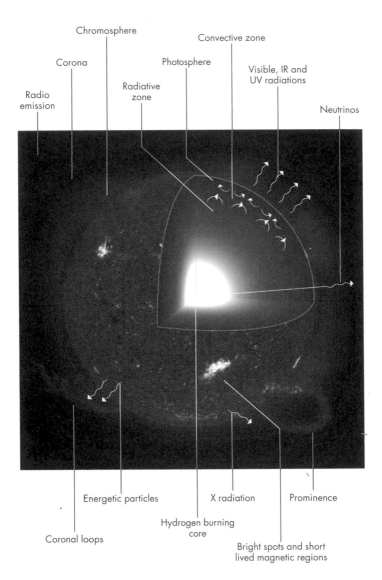

ABOVE Key features of the Sun.

FACT

Everything that is critical to your survival, from the food that you eat to the fossil fuels that you burn, gets its energy from sunlight. Even wind power relies on energy from the Sun to heat the atmosphere and create wind currents.

Death of a star

The life-giving energy that we receive from the Sun is released when hydrogen atoms in the core of the star fuse together to form helium. This process has been going on for a very long time – around five billion years in fact – and should continue for another five billion years, after which most of the hydrogen at the Sun's core will have been used up.

However, under the irresistible pressure of gravity, the core, no longer able to support itself, will begin to collapse as it begins the process of dying. Like a pressure cooker on the brink of blowing, this slow contraction will push up pressures and temperatures at the heart of the star.

Any remaining hydrogen will fuse to form helium in a shell around the core, releasing enormous amounts of energy. This will cause the outer layers of the Sun to swell horribly as it takes on the form a red giant. Soon our parent star, the supporter of all life on Earth, will consume both Mercury and Venus, stopping just short of the Earth where our fiery new neighbour will turn the surface of our planet into Hell's very own desert.

The core will become hot enough to cause the helium to fuse into carbon. When the helium fuel runs out, the core will expand and cool. The next stage is that the outer layers of our red giant sun will gradually drift off into space, forming what is referred to as a planetary nebula. The remaining core will then dim little by little, ending its days as a fading white-dwarf star where once it shone a bright light on civilisation.

The Moon

The origins of the Moon remain a mystery. The Earth's only natural satellite is an airless, lifeless, barren and dusty rock, its surface pitted by countless collisions with cosmic debris, and yet it remains the most arresting sight in the night sky.

ABOVE The Moon, showing its rocky surface.

The formation of the Moon

We can be fairly certain that the Earth and the Moon were formed at roughly the same time. Beyond that we can do little more than guess. The Moon may have formed alongside the Earth, making it a constant companion for the last four and half billion years or so. Alternatively, it may have become captured by the gravitational influence of the Earth at a later date, although this seems unlikely as the resulting heat would have melted both the Earth and the Moon.

The current most popular theory is that the Moon formed when the Earth was hit by a Mars-sized asteroid, a mighty collision that sent enough matter out into orbit around the Earth for the Moon to form from the resulting debris.

Phases of the Moon

For those who don't already know, the Moon orbits the Earth once every month. This ash-grey rock is visible on Earth whenever light from the Sun is reflected by its surface. Depending on the positions of the Earth and the Moon in relation to the Sun, the amount of the Moon's surface we can see in the night sky varies. These variations, during which the Moon seems to disappear before reappearing as a slim crescent that gradually grows to a full circle, are known as the phases of the Moon.

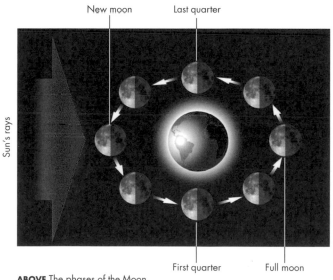

ABOVE The phases of the Moon.

The surface of the Moon

Pitted, scarred and dusty, the surface of the Moon has changed little in billions of years aside from suffering the obvious effects of meteorite and other impacts. Acting as a shield against certain death, the Moon has caught many an extinction-level meteorite or asteroid before it travelled that last quarter of a million miles or so and wiped out life as we know it. Up to 300 km (187 miles) across, the resulting craters can still be seen easily from Earth, highlighted against the surface of the Moon by huge, bright rays of ejecta which earn these battle scars the name 'ray craters'.

BELOW The Tycho crater, a ray crater on the near side of the Moon.

Eclipses

An eclipse is a natural occurrence, the result of the constant cosmic ballet performed by bodies in motion through space. A solar eclipse occurs when the Moon passes between us and the Sun, blocking the light from our parent star and casting a shadow over part of the surface of our planet. A lunar eclipse can be seen when the Earth passes between the full Moon and the Sun, casting a shadow over the surface of the Moon.

Solar eclipses occur once or twice a year and are only visible in the area of shadow cast by the Moon (This shadow is around 160 km/100 miles wide). Lunar eclipses occur two or three times a year and can be seen from any point on the Earth facing the full Moon.

A total solar eclipse is probably a relatively rare event anywhere in the galaxy. We have the opportunity to see them here on Earth because our Sun, although 400 times wider than the Moon, is also 400 times further away from us than the Moon. This remarkable coincidence means that they both appear to be roughly the same size in the sky. This means that when the Moon passes in front of the Sun, for the observer on Earth it blocks out the light from the star.

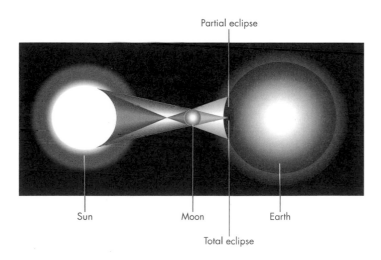

Partial eclipse

Sun   Moon   Earth

Total eclipse

ABOVE The Moon casts a shadow on Earth during a solar eclipse.

Comets

Occasionally called 'God's snowballs', comets originate in the coldest and darkest region of the Solar System. Hanging loose beyond the orbit of Pluto, these dirty snowballs, which are little more than lumps of rock and dust held together by ice and frozen gas, are occasionally knocked on to an elliptical orbit around the Sun by the gravitational influence of a passing star. Depending on the path it follows the comet may become caught in a predictable orbit and be recognised as a regular visitor to our part of space. Alternatively, it may pass just once by the Sun and head off into space from where it is unlikely to return during our lifetime.

Comets produce no light of their own but instead reflect light from the Sun. As a comet gets closer to the Sun its crust begins to melt, producing a cloud of gas and dust which is known as the coma or 'head' of the comet. A tail of gas is also released which, because it is deflected by the solar wind, always points away from the Sun. Although comets may be as large as 10 km (6.2 miles) in diameter, each pass around the Sun causes the comet to lose matter until eventually the giant snowball melts and disappears for ever.

In pre-scientific times, comets were often seen as harbingers of doom, a blazing message written large across the sky by the gods (or God). Sometimes, and purely by coincidence, the arrival of a comet did coincide with a significant event. This is hardly surprising given the frequency with which some of these comets return to our skies.

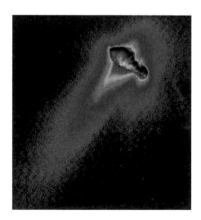

RIGHT The Borrelly Comet, showing its tail across the sky.

Major Comets	Discovered	Discoverer	Last (recorded) visit	Next visit	Period (years)
Halley	240 BCE	First sighted in antiquity	1986	2061	76
Kirch's	1680	Gottfried Kirch	1681	11036	9,355
Biela	1772	Wilhelm von Biela (1826)	1852	broke up in 1852	Not available
Encke	1786	Johanne Encke	2003	2007	3.28
Faye	1843	Hervé-Auguste-Etienne-Albans Faye	1999	2006	7.34
Swift-Tuttle, aka Kegler (responsible for the Perseids meteor shower)	1862	Ernst Wilhelm Leberecht Tempel and Lewis Swift	1992	2126	About 130
Wolf	1884	Max Wolf	2000	2009	8.21
Daylight Comet	1910	Multiple observers	1910	4144877	4,141,057
Arend-Roland	1957	Silvain Arend and Georges Roland	1957	unknown	unknown
Gehrels (78P)	1973	Tom Gehrels	1998/9	2012	5.5
Kohoutek	1973	Lubos Kohoutek	1973	76973	75,000
Howell (88P)	1981	Ellen Howell	1997	2009	7.2
Shoemaker-Levy 9	1993	Eugene and Carolyn Shoemaker and David Levy	1994	impacted Jupiter	Not available
Hale-Bopp	1995	Alan Hale and Thomas Bopp	None	4377	2,380
Hyakutake	1996	Yuji Hyakutake	1996	31496	29,500
2001 Q4 (NEAT)	2001	NEAT comet search program	2004	unknown	unknown
2002 T7 (LINEAR)	2002	LINEAR comet search program	None	unknown	unknown

The planets of our Solar System

Mercury

The closest planet to the Sun is the second smallest in the Solar System. The scarred and cratered surface covers a huge iron core, which accounts for 80 per cent of the planet's mass. The surface gravity of Mercury is too weak to hold on to a significant atmosphere, leading to an extremely broad range of temperatures during a single rotation.

Average distance from Sun: *57.91 million km (36.19 million miles)*
Diameter at equator: *4,879 km (3,049 miles)*
Orbital period: *87.97 Earth days*
Surface temperature: *-184°C to 450°C (-299°F to 842°F)*

Venus

The second rock from the Sun is a giant greenhouse of a planet shrouded in dense yellow clouds containing sulphuric acid. With atmospheric pressures 90 times those on Earth and surface temperatures even higher than those of Mercury, Venus presents an extremely hostile environment. Once thought of as a potentially habitable sister planet to Earth, Venus was for many years the ultimate destination of choice for the Soviet space program.

Average distance from Sun: *108 million km (67.50 million miles)*
Diameter at equator: *12,104 km (7,565 miles)*
Orbital period: *224.70 Earth days*
Surface temperature: *464°C (867°F)*

Earth

Home to the only life in the Universe that we know about, the Earth is the largest of the four inner rocky planets and the only one in the solar system to have an oxygen-rich atmosphere and liquid water. Life on the planet is protected from harsh solar radiation by a magnetosphere that is generated at the planet's core and which extends far out into space.

Average distance from Sun: *149.60 million km (93.50 million miles)*
Diameter at equator: *12,756 km (7,926 miles)*
Orbital period: *365.26 days*
Surface temperature: *-55°C to 70°C (-67°F to 158°F)*

The Moon

The daily tides on Earth are a result of the interaction of the gravitational forces between the Earth, the Moon and the Sun. Particularly high tides occur during a new moon or at full moon when the Sun and Moon are in line with the Earth. Low tides result from the Sun and Moon being at right angles to the Earth. About 384,000 km (239,000 miles) from Earth, the Moon completes its orbit every 27.32 days.

The far side of the Moon remained a mystery until October 1959 when a Soviet spacecraft sent back pictures of the 'dark side'. The Earth and the Moon are in synchronous rotation, which means that we always get the same view of the surface.

Mars

Famously known as the 'Red Planet' on account of the reddish iron-oxide dust that covers its surface, Mars is the planet that most resembles our own. It has clouds, canyons, mountains, valleys and deserts and even white polar caps. Despite this Mars is a cold, dry and lifeless planet with a thin atmosphere that is composed mostly of carbon dioxide.

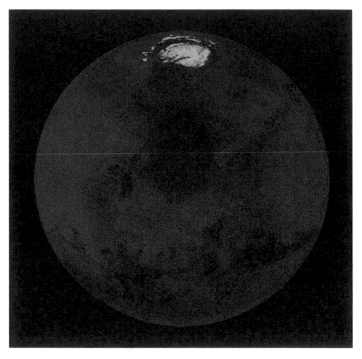

Average distance from Sun: *227.94 million km (141.64 million miles)*
Diameter at equator: *6,786 km (4,241 miles)*
Orbital period: *686.98 Earth days*
Surface temperature: *-120°C to 25°C (-184°F to 77°F)*

Jupiter

Mighty Jupiter is the king of the planets. A vast body composed mostly of hydrogen, it is in essence a failed star with its own mini-planetary system in orbit around it. The first of the four gas giants, Jupiter is 1,300 times bigger than the Earth and has a mass equal to two and a half times that of all the other planets combined.

Average distance from Sun: *778.29 million km (486.43 million miles)*
Diameter at equator: *142,984 km (88,846 miles)*
Orbital period: *11.86 Earth years*
Surface temperature: *-150°C (-184°F) (at cloud top)*

The moons of Jupiter

Jupiter has 16 moons, the most important of which are known as the Galileans, named for Galileo Galilei, who claimed to have discovered them when he looked at the heavens through his telescope in 1609. In fact, it was Simon Marius who first discovered them. These moons are similar in size to the Earth's moon and offer some of the most interesting sights in the solar system. They are the four biggest moons in the Solar System.

Io – the first of the Galilean moons of Jupiter is fiery Io. Its proximity to Jupiter means that it is constantly being churned by the gravitational influence of this failed star. This drives volcanic activity in an otherwise frozen region of space and leaves it with a surface of vivid reds and oranges which are believed to be evidence of the presence of various sulphur compounds.

Europa – the scientific community is very excited about Europa. Although clearly hostile to humans, this ice-covered ocean planet may just be home to life. There is the hope that volcanic activity deep below the ice crust may have resulted in a liquid sea with the possibility that life forms similar to those found around deep-ocean volcanic vents on Earth may also have evolved here.

ABOVE Io.

ABOVE Europa.

Ganymede – larger than either Mercury, Pluto or the Moon, Ganymede has a diameter of about 5,268 km (3,400 miles). The moon is composed of roughly equal amounts of rock and ice. It is scarred with many impact craters and, unusually, its surface features long parallel 'trenches', that are hundreds of metres deep and hundreds of kilometres long.

Callisto – the outermost of the four Galilean moons is pitted with impact craters which shine out from an otherwise dark and fairly featureless surface. Despite its relatively large diameter (4,800 km/ 3,000 miles), Callisto is only half as dense as the Moon. Callisto has the largest-known impact crater in the Solar System, Valhalla.

FACT

Powerful winds dominate the atmosphere of Jupiter with criss-crossing jet streams, lightning and huge hurricane-like storms, such as the Great Red Spot. This particular storm has been raging for over 300 years and is about twice the size of the Earth's diameter.

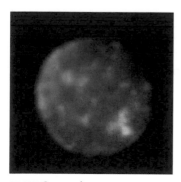

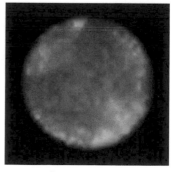

ABOVE Ganymede.

ABOVE Callisto.

Saturn

Famous for its magnificent and distinctive rings that were thought to be unique until 1977, Saturn is the least dense planet in the Solar System having a density that is less than that of water. It has 34 moons, and 13 recently discovered satellites and 7 main rings, 3 of which are visible from Earth through an ordinary telescope. The rings are composed of icy dust particles and ice-covered rocks that can be as small as a grain of sand or as large as a house.

Average distance from Sun: *1.43 billion km (893.75 million miles)*
Diameter at equator: *120,536 km (74,898 miles)*
Orbital period: *29.46 Earth years*
Surface temperature: *-180°C (-292°F) (at cloud top)*

The moons of Saturn

Saturn has many moons but the first to be discovered was Titan, spotted by the Dutch scientist Christiaan Huygens in 1656. The second-largest moon in the Solar System (after Ganymede), Titan is also one of only three known to have atmospheres.

The bulk of this moon is made up of rock and ice but it is the atmosphere of Titan that has attracted the interest of Earth-based scientists. Rich in nitrogen and other chemicals, it was the relatively recent discovery of molecules of hydrogen cyanide in Titan's atmosphere that really got their attention.

Since hydrogen cyanide is one of the key organic chemicals associated with the development of life on Earth, some scientists have speculated that this moon may one day be home to some form of life. Titan's surface temperature of -180°C (-292°F) means, however, that in the unlikely event of life evolving there it would almost certainly not resemble life as we know it.

Most likely locations for life

The most likely locations for life in our Universe, other than on Earth, are:

1. Mars: fossils of bacteria may exist all over the planet. Current life might be found deep below the surface, where liquid water may exist. Certainly, bacteria exist in similar conditions on Earth.

2. Europa: one of Jupiter's moons: may have liquid water beneath its icy crust, while strong tides and magnetic activity could generate heat. The combination could produce conditions similar to those near Earth's mid-ocean vents.

3. Titan: Saturn's largest moon: has a smog-like atmosphere rich in organic chemicals similar to Earth's early atmosphere; it also has frozen water and a hot core.

Uranus

An almost featureless planet lying in the icy depths of the outer Solar System, Uranus has a rocky core which is surrounded by a mantle of gases and ice. The planet's methane-rich atmosphere gives it a blue-green colour which, aside from its severe axial tilt (97.77° to the orbital axial plane) and striking ring system, is the only distinctive feature of the planet.

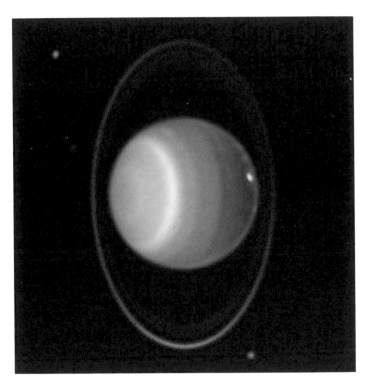

Average distance from Sun: *2.87 billion km (1.79 billion miles)*
Diameter at equator: *51,118 km (31,763 miles)*
Orbital period: *84.01 Earth years*
Surface temperature: *-210°C (-346°F)(at cloud top)*

The moons of Uranus

Uranus is known to have 27 moons, all of which are made of a dark and dusty mixture of rock and ice. The four largest are Oberon, Titania, Umbriel and Ariel. These are relatively dull worlds displaying the usual impact crater damage one would associate with bodies of this nature. Miranda, the innermost of the major moons of Uranus, is an entirely different matter.

Smaller than the other major moons, Miranda was discovered in 1948 by American astronomer GP Kuiper. Its surface possesses a riot of interesting features, including enormous cliffs, wide open plains and deep canyons, some of which are up to 10 times deeper than the Grand Canyon here on Earth; these canyons appear on photographs as deep grooves in the surface of the moon.

Scientists have speculated on how these unlikely features may have formed without the usual forces of erosion. The most popular explanation is that Miranda may at some time in the past have been shattered by a collision with a large object only to reassemble itself later into its current rather haphazard form. It is perhaps as a result of these collisions that Miranda, which is thought to be made up of ice and rock in a roughly 50/50 ratio, is also home to a cliff face that is an incredible 5 km (3.5 miles) high.

As with all the moons of Uranus, Miranda is named after a character from a Shakespeare play, in this case the daughter of the magician Prospero in *The Tempest*.

FACT

Uranus was the first planet to be discovered in modern times as it is not visible with the naked eye. It was spotted by William Herschel, while systematically searching the sky with his telescope on 13 March, 1781. It had been seen through telescopes before but was thought to be just another star.

Neptune

Similar in size and structure to its neighbour Uranus, Neptune is possibly the last 'true' planet in the Solar System. The atmosphere, which is bright blue in colour due to the presence of methane, is actually made up mostly of hydrogen. Like the other gas giants Neptune has a ring system, which is divided into four distinct bands.

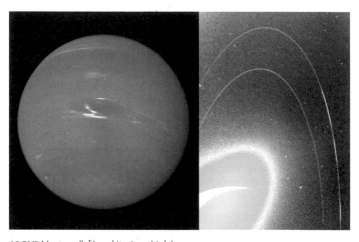

ABOVE Neptune (left) and its rings (right).

Average distance from Sun: *4.49 billion km (2.81 billion miles)*
Diameter at equator: *49,528 km (30,955 miles)*
Orbital period: *164.79 Earth years*
Surface temperature: *-220°C (-364°F) (at cloud top)*

FACT
The coldest place in the Solar System is Neptune's moon Triton where temperatures as low as -235°C (-391°F) have been recorded.

FACT

NASA's *New Horizons* spacecraft was launched aboard an *Atlas V* rocket in January 2006 and should reach Pluto in mid-2015. It will give us our first close-up look at Pluto and its moon, Charon, since no spacecraft has ever visited the planet and detailed images are beyond the scope of the Hubble Space Telescope.

Pluto

Believed by some astronomers to be nothing more than a truly huge asteroid, icy Pluto exists near the edge of the Solar System. It is so far from the Sun that our star would appear as little more than yet another bright light in the skies over Pluto. Its moon, Charon, is about half the size of Pluto and was only discovered in 1978.

ABOVE Pluto and its moon, Charon.

Average distance from Sun: *5.91 billion km (3.69 billion miles)*
Diameter at equator: *2,390 km (1,431 miles)*
Orbital period: *248.54 Earth years*
Surface temperature: *-230°C (-382°F)*

How much would I weigh on . . . ?

We have evolved here on Earth under the influence of gravity. The gravitational pull that the Earth exerts on us is largely determined by the mass of this planet. Your weight is a consequence of the interaction of gravity with mass. However, on another planet, one possessing a different mass, your weight would be different from that on Earth.

The following chart represents an interplanetary guide to weight loss (and gain) and is based on a starting weight, on Earth, of 64 kg (10 st 11 lb).

Planet	Relative Mass (Earth = 1)	Resulting Weight
Mercury	0.06	24 kg (3 st 11 lb)
Venus	0.82	57 kg (9 st)
Earth	1.0	64 kg (10 st 1 lb)
The Moon	0.01	10 kg (1 st 8 lb)
Mars	0.11	24 kg (3 st 11 lb)
Jupiter	317.83	150 kg (23 st 9 lb)
Saturn	95.16	58 kg (9 st 2 lb)
Uranus	14.53	56 kg (8 st 11 lb)
Neptune	17.14	71 kg (11 st 3 lb)
Pluto	0.0021	4 kg (9 lb)

FACT
One of the first people to find a practical use for Newton's newfangled gravity equations was Edmond Halley, who used them to show that the comets seen in 1531, 1607 and 1682 were in fact the same comet, which would be returning in 1758. It did, and is now called Halley's Comet.

STARS AND
CONSTELLATIONS

What is a star?

There is no such thing as a 'standard' star. These enormous, hot, luminous balls of gas vary enormously depending on their mass, a factor which affects everything from the star's size, temperature and colour through to its brightness and the length of its life. Even with apparently similar stars there are variations in their characteristics because of internal changes which occur during the lifetime of a star.

Most of the visible matter in the Universe is contained within stars, of which there are believed to be trillions in the Universe. (Our own Milky Way Galaxy contains billions.) The closest star to the Earth (aside, obviously, from the Sun) is Proxima Centauri, which lies a little over four light years away. The furthest stars from us can be seen in galaxies which are situated at the limit of our technology, glimpses of which can be captured by effectively looking billions of years back in time to the moment when the light from these stars began its journey to Earth.

In truth, it's highly likely that these stars are no longer there; what we see is a brief snapshot of cosmic history preserved as a consequence of the time taken for the light from these stars to reach us.

The Sun is unusual in so far as it is a true lone star. Far more common are pairs of stars, multiple systems or clusters of stars consisting of several components including nebulae (which are clouds of gas and dust). These stars orbit a common centre of gravity which usually lies at a point beyond the perimeter of any individual star. Often, in binary (double) star systems one of the stars will be too faint to be seen from Earth. In these cases astronomers may still be able to identify a binary system from the effects that the invisible star may have on its companion, affecting its motion and also impeding the light coming from the star by passing between it and the observer here on Earth.

RIGHT Star clusters in the Large Magellanic Cloud, taken by Hubble Space Telescope.

Life cycle of a star

It's been said that death is the only thing in life of which one can be certain, and this is true even for stars. They are born, grow to maturity, gradually fade away and then finally die. Death can come in many forms, some spectacular shows, others rather slow and sad, but it is certainly one of the things that we humans share with our cosmic companions.

Stars are born in huge clouds of gas and dust, the scale of which is almost beyond our comprehension. The Eagle Nebula is just such a huge cloud of gas and dust with tiny finger-like protrusions sticking out from the very top of one of its pillars of gas. The distance across a single one of these 'fingertips' is equal to the distance from our Sun to Pluto, which is around 5.91 billion km (3.69 billion miles). Amazingly, any one of these fingertips could easily accommodate our entire solar system.

Star formation begins when the cloud of gas and dust is disturbed, an event which triggers an increase in density in a region of the cloud. Gravity causes the dense region to contract, pulling in more matter and making it evermore dense. At this point it becomes ball-shaped. From a starting point at just a few degrees above Absolute Zero (-273.15°C/-460°F), the cloud begins to heat up and then starts to glow. Now spinning at great speed, the protostar, a star in the making, begins to give out enough light and heat for it to be detected using an infrared telescope.

FACT
Stars cannot grow indefinitely and nature places a limit on the maximum size that a star can grow to. Anything more than 120 times larger than our Sun will be destroyed by its own radiation.

By now the original cloud of gas and dust has condensed at the core of the star and its temperature will be rising rapidly until the point, at around 10 million°C (18 million°F), when it bursts into life as hydrogen nuclei combine to form helium, releasing enormous amounts of energy in a process known as nuclear fusion. From this point on the life cycle of the star will largely be determined by its mass (the amount of matter it contains).

Low mass star

A low mass star, one with perhaps less than a tenth of the mass of our Sun, will by comparison give off a feeble glow yet burn for tens of billions of years. Called a red dwarf, it gets its name from the colour of the light it produces, which indicates that its surface temperature is relatively low, usually around 3,500°C (6,300°F). Red dwarf stars are the most common type of star in the observable Universe.

ABOVE Red dwarf: artist's impression of gas giant planet orbiting the red dwarf star Gliese 876 in the autumn constellation Aquarius.

Medium mass star

A star of medium mass, which is to say somewhere between a red dwarf and perhaps half as big again as our own Sun, has a shorter but far more interesting life. This type of star is of particular interest to us on Earth as our own Sun falls into this category and its fate (and our own) is shared with other stars of similar mass.

Having moved from the protostar stage to the main sequence, which is the longest portion of any star's life, these stars will usually shine brightly for around 10 billion years. (Our Sun is around half-way through its main-sequence stage). Once most of the hydrogen fuel has been used up, the outer layers of the star initially begin to collapse inward until the effects of the corresponding rising pressure and heat cause the star to expand again and become a red giant.

While this occurs, elements such as carbon and helium begin to 'burn' at the core of the star as the outer layers gradually drift off into space, creating what astronomers call a planetary nebula. Before long only the dense core remains, shrinking ever smaller and getting gradually dimmer until all that remains is a white dwarf, a cosmic ghost of the star that once was.

Large mass star

Stars with a mass at least half as large again as our Sun's live short but brilliant lives. They burn their fuel rapidly over millions rather than billions of years before expanding into red supergiants. When the core of a red supergiant collapses it does so with explosive consequences. This explosion, called a supernova, blasts off the outer layers of the star producing a dramatic flash that can be seen from millions of light years away. Should the core of the star survive the explosion it will cool and contract further to form a small and incredibly dense neutron star or a pulsar. A neutron star is composed either entirely or mostly from neutrons and gives off comparatively little light. A pulsar is a rotating neutron star which emits a strong and regular burst of radio waves as it rotates.

Super mass star

Stars with a mass in excess of 100 times that of our Sun are called supergiants. These magnificent creatures shine up to a million times brighter than our Sun yet pay a heavy price for their brilliance. After just a few million years all of their fuel is burned up and the star collapses under its own immense gravity. When this process begins it will not stop and as the core grows evermore dense even light itself cannot escape. What was once among the brightest objects in the Universe now becomes that darkest, most intriguing and exotic of objects – a black hole.

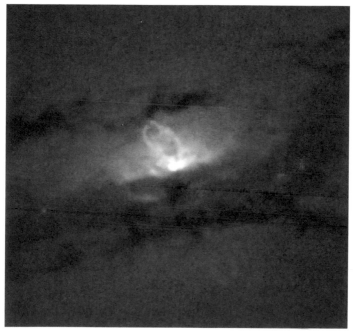

ABOVE A black hole blowing bubbles.

FACT
Stars do not stay in the same position in the night sky – they rise in the east and set in the west in the same way that our Sun does. If you are looking at the stars through a telescope you will see that they are constantly moving and so would produce blurred star trails if left pointing in the same direction for hours. For this reason astronomers mount their telescopes on special supports that rotate at the same speed and in the same direction as the stars overhead.

Element factories

Back in 1932 the German physicist Werner Heisenberg was awarded the Nobel Prize for his work on Quantum Mechanics, putting forward the theory that all elements could be built up from protons and electrons, which is a roundabout way of saying that they could all be made from hydrogen, the raw fuel of stars.

In 1938 another German physicist, named Hans Bethe, provided a satisfactory explanation for how heavier elements could also be built up this way. In short, he suggested that collapsing stars act as element factories, with elements being built up from protons and electrons at the core of the star.

Following on from this idea, the Indian-born astrophysicist Subrahmanyan Chandrasekhar examined the final stages in the life of truly massive stars. He suggested that as a star of this type explodes in a violent supernova, the explosion will synthesise heavy elements such as carbon, iron and even gold, blasting them across vast distances in space. Over time, these elements will provide the raw matter for everything in the Universe, from new stars and planets to plants and living organisms.

As a carbon-based life form, the elements from exploding stars are part of your physical make-up. In essence, we were all once simply stardust . . . and probably will be again.

Distances to the stars

The Universe, as we've established, is so big that words cannot really convey a sense of the scale of the thing. Distance measured in kilometres means nothing beyond the boundaries of our solar system so instead we use nature's own yardstick: light. A light year is the distance travelled by a beam of light across the vacuum of space in one year. Light travels about 300,000 km (186,000 miles) in one second, a remarkable 9.461 trillion km (5.913 trillion miles) in a single year.

Here is a three-dimensional view of the constellation of Orion. From Earth, the stars that make up the hunter of mythology appear as a coherent flat image in the night sky but they are in fact separated by vast distances. Each of the divisions in the illustration is equal to five hundred light years, the distance that light will travel in five hundred years.

ABOVE A view of the constellation of Orion.

The constellations

The 88 constellations are given here with the hemisphere in which they are primarily visible (North or South). Constellations become higher or lower in the sky with the changing of the seasons. Thus many of them will be visible in different hemispheres at different

	Constellation	Hemisphere most visible		Constellation	Hemisphere most visible
1	Andromeda	N	24	Columba (The Dove)	S
2	Antlia (The Air Pump)	S	25	Coma Berenices (Berenice's Hair)	N
3	Apus (Bird of Paradise)	S			
4	Aquarius (Water Carrier)	N/S	26	Corona Australis (Southern Crown)	S
5	Aquila (The Eagle)	N/S	27	Corona Borealis (Northern Crown)	N
6	Ara (The Altar)	S			
7	Aries (The Ram)	N/S	28	Corvus (The Crow)	N/S
8	Auriga (The Charioteer)	N	29	Crater (The Goblet)	N/S
9	Boötes (The Herdsman)	N	30	Crux (Southern Cross)	S
10	Caelum (The Chisel)	S	31	Cygnus (The Swan)	N
11	Camelopardalis (The Giraffe)	N	32	Delphinus (The Dolphin)	N/S
12	Cancer (The Crab)	N/S	33	Dorado (The Goldfish)	S
13	Canes Venatici (Hunting Dogs)	N	34	Draco (The Dragon)	N
			35	Equuleus (The Little Horse)	N/S
14	Canis Major (The Greater Dog)	N/S	36	Eridanus (The River)	N/S
15	Canis Minor (The Lesser Dog)	N/S	37	Fornax (The Furnace)	S
16	Capricornus (The Goat-Fish)	N/S	38	Gemini (The Twins)	N/S
17	Carina (The Keel)	S	39	Grus (The Crane)	S
18	Cassiopeia	N	40	Hercules	N
19	Centaurus (The Centaur)	N/S	41	Horologium (The Pendulum Clock)	S
20	Cepheus	N			
21	Cetus (The Whale)	N/S	42	Hydra (The Water Serpent)	N/S
22	Chamaeleon	S	43	Hydrus (The Lesser Water Snake)	S
23	Circinus (The Drafting Compass)	S	44	Indus (The Indian)	S
			45	Lacerta (The Lizard)	N

times of the year, while some will be visible to both hemispheres for much of the time.

Where a star is visible in both hemispheres for significant parts of the year, it is marked N/S on this chart.

	Constellation	Hemisphere most visible		Constellation	Hemisphere most visible
46	Leo (The Lion)	N/S	70	Reticulum (The Reticle)	S
47	Leo Minor (The Lesser Lion)	N	71	Sagitta (The Arrow)	N
48	Lepus (The Hare)	S	72	Sagittarius (The Archer)	N/S
49	Libra (The Scales)	N/S	73	Scorpius (The Scorpion)	N/S
50	Lupus (The Wolf)	N/S	74	Sculptor (The Sculptor)	S
51	Lynx	N	75	Scutum (The Shield)	N/S
52	Lyra (The Lyre)	N	76	Serpens (The Serpent)	N/S
53	Mensa (The Table Mountain)	S	77	Sextans (The Sextant)	N/S
54	Microscopium (The Microscope)	S	78	Taurus (The Bull)	N/S
			79	Telescopium (The Telescope)	S
55	Monoceros (The Unicorn)	N/S	80	Triangulum (The Triangle)	N
56	Musca (The Fly)	S	81	Triangulum Australe (Southern Triangle)	S
57	Norma (The Level)	S			
58	Octans (The Octant)	S	82	Tucana (The Toucan)	S
59	Ophiuchus (Serpent Holder)	N/S	83	Ursa Major (The Greater Bear)	N
60	Orion (The Hunter)	N/S	84	Ursa Minor (The Lesser Bear)	N
61	Pavo (The Peacock)	S	85	Vela (The Sails)	S
62	Pegasus (The Winged Horse)	N	86	Virgo (The Virgin)	N/S
63	Perseus	N	87	Volans (The Flying Fish)	S
64	Phoenix (The Phoenix)	S	88	Vulpecula (The Fox)	N
65	Pictor (The Painter's Easel)	S			
66	Pisces (The Fish)	N/S			
67	Piscis Austrinus (The Southern Fish)	S			
68	Puppis (The Stern)	S			
69	Pyxis (The Compass)	S			

Star map

Here are the major constellations as they appear in the night sky. The constellations that you will be able to see will depend on your position on the planet. For the sake of convenience, we have divided the night sky into the northern and southern hemispheres. The positions of the constellations appear to change depending on the time of year with the result that some apparently drop below the horizon only to reappear later.

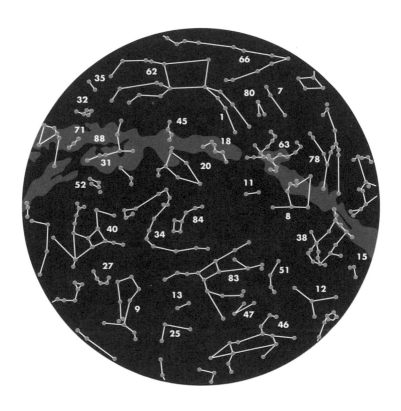

Finding a constellation in the night sky can be a tricky business so it is best to find a reference point that you can recognise – such as the Pole Star, when observing in the northern hemisphere – and then find your way about from there.

BELOW Constellation maps of the northern and southern celestial spheres. Match the numbers to the numbers on the previous pages to see their names.

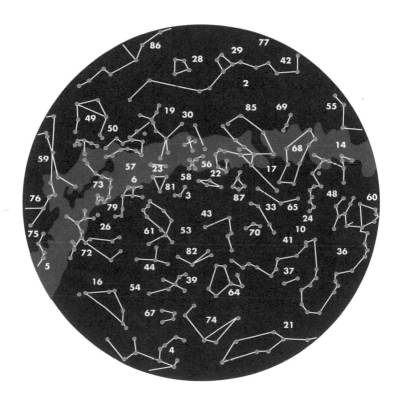

The zodiac

From our perspective here on Earth, the Sun appears to move against the background of the stars. The path that the Sun follows is known as the ecliptic, and a narrow band extending to about 9° either side of the ecliptic is called the zodiac.

In times past the zodiac was divided into two constellations, through which the Sun passed in the course of a single year. Originally intended as a way of measuring the passage of time, all manner of superstitions developed in association with these constellations, with astrology growing out of the belief that the positions of the stars and planets at the time of birth will influence the outcome of one's life.

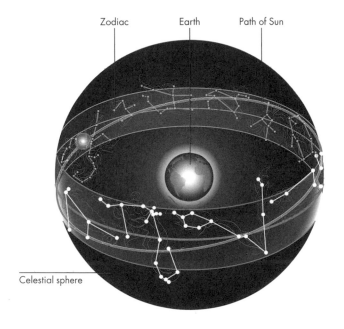

Zodiac   Earth   Path of Sun

Celestial sphere

ABOVE The 12 astrological constellations in a band around the Sun.

Most of the constellations owe their names to the ancient Greeks, who clearly had very vivid imaginations. Below is an example of an ancient constellation and the mythological figure imposed upon it.

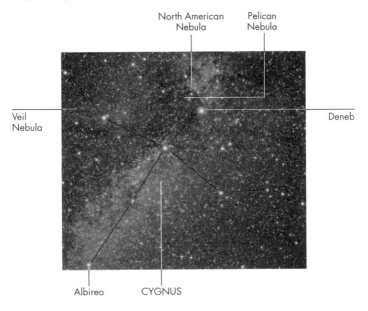

North American Nebula

Pelican Nebula

Veil Nebula

Deneb

Albireo

CYGNUS

ABOVE Cygnus, one of the richest constellations in the sky. Ancient astronomers believed it formed the shape of a swan with Albireo as the head and Deneb as the tail.

FACT

When astronomers pointed out recently that there are in fact 13 constellations in the zodiac – the 13th being Ophiuchus, the Serpent Bearer, which lies between Scorpio and Sagittarius – they were accused by astrologers of being 'unscientific'.

Star magnitudes

The brightness of a star, which is determined by the amount of radiation it emits, is measured in magnitudes. Confusingly, the brightest stars are given the lowest figures of magnitude. To make matters worse, there are two different kinds of magnitude: apparent and absolute.

Apparent magnitude is a measure of the brightness of a star as seen from Earth. Inevitably, this means that the stars nearest to us tend to appear brighter, while the light coming from those further away will spread out as it travels to us through space, making them appear less bright.

Absolute magnitude is a more realistic measure of a star's brightness. This figure is based on the magnitude of a star as observed from a standard distance of 32.6 light years away (10 parsecs). This removes the subjective element from the observation and is a far more accurate measure of the true nature of the star.

The Twenty Brightest Stars

Name	Apparent Magnitude	Absolute Magnitude	Name	Apparent Magnitude	Absolute Magnitude
Sirius	-1.43	1.4	Hadar	0.61	-5.42
Canopus	-0.72	-5.63	Acrux Altair	0.77	2.21
Rigil Kentaurus	-0.01	4.34	Altair Aldebaran	0.85	-0.65
Arcturus	-0.04	-0.3	Aldebaran Acrux	0.86	-0.8
Vega	0.03	0.58	Spica Antares	0.96	-5.38
Capella	0.08	-0.48	Antares Spica	0.98	-3.55
Rigel	0.12	-6.75	Fomalhaut Pollux	1.14	1.07
Procyon	0.38	2.66	Pollux Fomalhaut	1.16	1.73
Betelgeuse Achernar	0.46	-2.76	Deneb	1.25	-8.73
Achernar Betelgeuse	0.50	-5.09	Mimosa	1.25	-3.92

FACT

Twinkling stars twinkle because of refraction caused by the movement of air in the Earth's atmosphere. In reality, the light coming from the star remains constant.

Starlight

Stars are categorised according to their spectral characteristics, i.e., according to the intensity of the radiation they emit. Starlight can be analysed by passing it through a prism or, more likely these days, through a device called a diffraction grating.

This produces a spectrum of colours with dark lines which indicate the presence of atoms of certain elements in the star. This provides astronomers with what amounts to a list of ingredients for the star and is also very useful for determining the nature and probable life cycle of any given star.

Once analysed, stars are categorised into one of seven main spectral classes, each of which has been given a letter of the alphabet: O, B, A, F, G, K, M. These seven classes are further divided into sub-classes numbered 0 to 9. The Sun belongs to the spectral class G2.

FACT

Hipparchus, the ancient Greek astronomer who realised that the Earth is not flat, gave us our orders of magnitude when he described the 20 brightest stars he could see as being of the 'first' magnitude while the least bright were said to be of the 'sixth' magnitude.

Classifying stars

One can tell a great deal about a star from its colour, but this can be deceptive. We often talk about something being 'red hot' when what we really mean is 'extremely hot'. As ever, these words mean little when applied to the environment of outer space. The hottest stars are actually blue in colour, with the 'coolest' (e.g., those around 3,500°C/3,600°F to 5,000°C/9,000°F) being red.

This relationship between colour and luminosity (brightness) can be plotted on a graph known as the Hertzsprung-Russell diagram. This is a simple device which plots a star's spectral type against its absolute magnitude. The hottest stars are plotted toward the left-hand

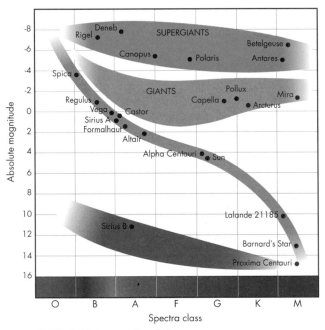

ABOVE The Hertzsprung-Russell diagram.

side of the diagram with the coolest to the right. Similarly, the brightest stars are plotted near the top of the diagram while the dimmest are placed near the bottom.

Once plotted, this relationship between colour and temperature produces a broadly diagonal band across the middle of the diagram. Stars in this part of the graph are said to be main sequence stars, and this is where most stars are to be found for the majority of their lives. (Our Sun, for example, is classified according to the Hertzsprung-Russell diagram as a main sequence yellow dwarf.) The exceptions are the supergiants, which are found above the main sequence, and the tiny dwarf stars, which exist below it.

Star mass

Stars can be classified according to their mass. This figure is an expression of the amount of matter something contains, i.e., a measure of its density. When describing the mass of a star, we use our own Sun as a baseline. The Sun is considered to possess one solar mass. The dimmest dwarf stars may have a solar mass of only 0.08, while a supergiant may have a solar mass of 120. Most stars lie somewhere between 0.08 and 60 solar masses.

A star's mass (the amount of matter it contains) may be entirely unrelated to its volume (the amount of space it fills). Betelgeuse, for example, is a pulsating red–supergiant with an average diameter that is 650 times that of the Sun. This means that its volume is 40 million times greater than the Sun's volume. Despite this, Betelgeuse contains little more than 15 times the mass of the Sun, which means that even the air here on Earth is denser than the red supergiant Betelgeuse.

Type of star	Mass of a thimbleful	Equivalent on Earth
The Sun (main sequence yellow dwarf)	1.4 g (0.05 ounces)	Half a sugar lump
White dwarf	1.25 tonnes (1.38 tons)	A hippopotamus
Neutron star	approx. 179 million tonnes (197 million tons)	A fifth of the world's population

Star companions

Our star, the Sun, is unremarkable in many ways but one: it is a lone star. Around half of all recorded stars exist in tandem with a companion star in an arrangement known as a binary star system. Binary stars are locked into orbit around a common centre by the force of gravity and the time taken for them to orbit this point is known as the orbital period. So far orbital periods of between one day and a hundred years have been observed in binary star systems.

Eclipsing binary stars

The light coming from a distant star can appear to vary from one night to another. In many cases this is due to the presence of a smaller, unseen star in orbit around it. This type of arrangement is known as an eclipsing binary star system. As the smaller and dimmer star passes between the visible star and our observation point here on Earth, it causes a characteristic dip in the amount of light we can see. This periodic fluctuation in magnitude can be plotted on a graph. The resulting line is known as a light curve.

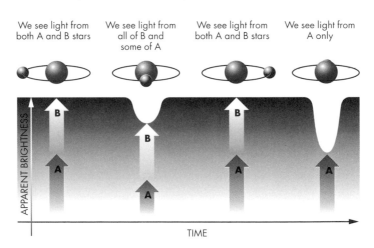

We see light from both A and B stars   We see light from all of B and some of A   We see light from both A and B stars   We see light from A only

APPARENT BRIGHTNESS

TIME

ABOVE The light perceived from a star with a smaller eclipsing binary in orbit around it.

Star clusters

Stars are often found in groups known as clusters. A star cluster is an arrangement of stars held together by mutual gravitation. These stars share a common history, having all been born from the same gigantic cloud of gas and dust. This means that they will be of a similar age and share the same chemical make-up. (They will differ from each other, however, because the real determinant of a star's character or type is its mass.) Star clusters divide roughly into two types: globular and open.

Globular clusters

Stars in a globular cluster will have formed when the Milky Way Galaxy was still little more than a spherical cloud of gas and dust. Globular clusters are found in or very near to the spherical halo which surrounds the centre of the galaxy. So far about 150 globular clusters have been spotted in this region of the galaxy, most containing stars that are in excess of 10 billion years old.

ABOVE Globular cluster M31.

FACT

On the off-chance that there might be someone listening, scientists sent a radio message from Earth to the globular cluster known as M13 back in 1973. Conveying information about our Solar System, human beings and DNA, it will not actually reach M13 until some time in the year 30,000.

Open clusters

As the name suggests, an open cluster is a very loose arrangement of stars. The number of stars in any one cluster may vary from less than a hundred up to several thousand. Stars in open clusters are newer than those found in globular clusters and are situated in the rotating disc or arms of the Milky Way Galaxy. These are young and very bright stars which may be destined to drift away from the rest of the cluster under the influence of gravitational disturbances from other objects in the galaxy.

ABOVE Pleiades (The Seven Sisters), a beautiful example of an open cluster.

Nebulae

A nebula is a cloud of gas and dust. Sometimes referred to as 'star nurseries', these clouds constitute some of the largest objects in the Universe and can provide all the necessary matter for star formation. Nebulae are classified according to their relationship with light and can be divided into two main types: bright nebulae and dark nebulae.

Bright nebulae

Associated with star birth, bright nebulae consist of two main types: emission and reflection. Emission nebulae glow in the red part of the spectrum because they are heated by the young stars in their midst. Reflection nebulae, on the other hand, tend to appear blue in colour because the dust they contain scatters the light coming from young stars in or around them.

There are two other types of bright nebulae, both of which occur as a result of the death of stars. First is the planetary nebula, which is created when a dying red giant throws off its gaseous outer layers to form an expanding shell of glowing gases. These nebulae, which are absolutely spectacular to look at, last for a few tens of thousands of years – a mere blink in cosmological time – before they disappear forever.

Secondly, a supernova remnant is a gas cloud that was born in the fierce explosion of a giant star. This casts enormous amounts of debris across space which is heated by the shock wave that results from the supernova explosion.

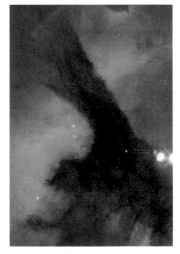

ABOVE The Trifid Nebula, an unusual example of a combined emission and reflection nebula.

ABOVE The Cygnus Loop (a supernova remnant).

Dark nebulae

Also known as absorption nebulae, dark nebulae are clouds of gas and dust which absorb the light passing through them and do not radiate any visible light – although they may radiate radio waves or infrared energy. If there is sufficient material in a dark nebula the process of star formation may begin and the nebula will undergo a change to become an emission nebula. Dark nebulae are sometimes difficult to detect and are best seen against a background of bright nebulae.

It should come as no surprise to discover that Messier's Catalogue, the original (relatively) modern account of all of the visible features of the night sky, contains not a single example of a dark nebula. Invisible to the naked eye (and all optical telescopes other than by inference), dark nebulae are usually noted for what they are believed to obscure. In recent years the most famous of the dark nebulae, the Horsehead Nebula, has become an almost iconic image, embodying the beautiful mystery of space and the Universe and presenting one of the last truly unknown and yet familiar regions of the heavens.

Death of a star

As a star dies, which is to say when the nuclear reactions at its core begin to shut down, the star becomes unstable. In the case of a star such as the Sun this triggers a process that causes the star to evolve fairly rapidly into a red giant before it is transformed into a feeble white dwarf engulfed in an expanding shell of gases known as a planetary nebula.

Beautiful as planetary nebulae are, when a massive star dies things get a lot more exciting – and much more spectacular.

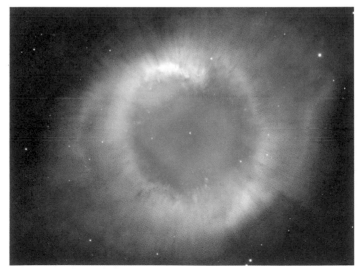

ABOVE The Helix Nebula, which is a planetary nebula.

Going supernova

A star must possess at least 10 solar masses to qualify as a supergiant. This means that it needs to have at least 10 times the mass of our own Sun. The life of a star that is this massive will end in an explosion of energy so powerful that it will, for a short time, outshine a galaxy of billions of stars.

Despite its size, a high mass star will use up its fuel over millions rather than billions of years. When it evolves into a supergiant the outer layers of the star will expand until its radius is roughly a thousand times that of our Sun. Inevitably, the star's gravity will cause its core to collapse. This happens at a tremendous speed – usually less than a second – triggering a correspondingly galaxy-shaking explosion which throws debris right across space. The star can now truly be said to have gone supernova.

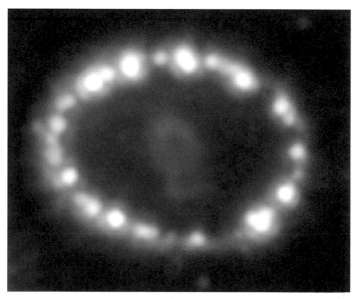

ABOVE Supernova 1987A, taken by Hubble Space Telescope.

Depending on the original mass of the supergiant, and the effects of the explosion, the remaining core of the star – if it still exists – will continue to evolve in one of two ways. It will either become a neutron star or a black hole.

Neutron stars and pulsars

If, after a supernova, the core of the star remains intact and enough matter remains for it to be between 1.4 and 3 solar masses then gravity will continue to do its job, compressing the star beyond the usual white-dwarf stage – which will be the fate of our Sun – and on into far more exotic territory.

Under extreme compression, the protons and electrons in the star's core will be crushed together to form neutrons. The process continues until the star, which you may recall was once a thousand times wider than our Sun, is little more than 20 km (12.5 miles) in diameter. The resulting object is called a neutron star.

A young neutron star spins rapidly on its axis and emits incredibly powerful beams of radiation that are channelled by the star's magnetic field into two powerful beams. Should one of these beams pass over the Earth it will be detected as a regular pulse of radio waves. Neutron stars that behave in this way are, quite logically, called pulsars. Pulsars can also give off other forms of electromagnetic radiation, such as intense X-rays and visible light.

The rate at which a neutron star pulses is determined by its rate of spin. A slow pulsar rotates perhaps once every four seconds; fast pulsars spin at a rate of about 30 times per second. Binary pulsars can spin at a rate of up to a thousand times a second.

FACT

In order for the Earth to become a black hole, it would have to be compressed until it had a diameter of just one centimetre.

Black holes

If after a supernova there remains a core that is in excess of three solar masses then events tend to drift over into the realms of what might almost be called science fiction. With a remaining core of over three solar masses gravity is able to overcome any pressure from the matter at the core and, in theory at least, continue to compress it to the point where it has zero volume but infinite density. The remaining entity is called a singularity and it is at this stage that the conventional laws of physics appear to break down.

The intensity of the gravitational field thrown up by a singularity is so great that it creates a region of space around itself where nothing, not even light, can escape. This region is known as a black hole. The boundary of a black hole, the point of no return, is known as an event horizon.

Exoplanets

Exoplanets are planets that are orbiting stars outside our own Solar System. The first such planet, 51 Pegasi A, was discovered in 1995 by a Swiss team led by Michel Mayor and Didier Queloz. Since then, over 150 exoplanets have been detected.

The oldest exoplanet: *Orbits a binary star in the M4 globular cluster and could be 13 billion years old.*

The most distant exoplanet: *The same as the M4 globular cluster, it is 5,600 light years away.*

Closest and smallest exoplanet: *Epsilon Eridani C, which is one-tenth as massive as Jupiter and 10 light years away.*

FACT

If a bag of sugar, which ordinarily weighs 1 kg (2.2 lb) here on Earth, were to be taken to the edge of a black hole, its weight would increase under the influence of such extreme gravity to a trillion metric tons.

THE
STUFF
OF THE
UNIVERSE

What is the Universe made of?

The Universe, which is to say everything that exists, is largely composed of matter surrounded by not very much at all. That said, the 'not very much at all' may yet turn out to include something rather important, so for now we call some of it 'dark matter'.

Most of the matter that we can see is arranged in enormous 'superclusters' of galaxies. Each of these galaxies is made up of billions of stars, some of which have planets in orbit around them. At least one of these planets is home to living creatures.

ABOVE A quintuplet cluster near the centre of the galaxy.

It is believed that all of the matter in the Universe came into being during the Big Bang. Currently the only credible model of the origin for the Universe, the Big Bang occurred around 14 billion years ago when the Universe spontaneously exploded into existence.

During the period immediately after the explosion all that existed was radiation and an assortment of subatomic particles. As the Universe expanded, some of this matter condensed to form the stars, which are concentrated in the galaxies which, as we already know, are themselves grouped into superclusters.

All of the matter contained in the Universe is composed of tiny particles, which are invisible to the naked eye. Protons, neutrons and electrons are examples of these particles and they are usually found grouped together in the form of atoms – an atom is just about the smallest thing an element can be broken down into and still retain its basic properties. Even these particles, however, are made of smaller particles called quarks, which can be one of six types or 'flavours': up; down; top; bottom; strange; and charm.

Antimatter

As the name suggests, antimatter might be considered to be the opposite of matter. Certainly, most of the attributes one would normally associate with elementary particles – such as electrical charge and spin – are reversed in particles of antimatter.

Particles of antimatter are known as positrons and they cannot exist in close proximity to particles of matter; the particles annihilate each other, releasing huge amounts of energy in accordance with Einstein's theory of special relativity.

Some astronomers have suggested that antimatter galaxies may exist in the far reaches of the Universe, although others argue that taking such a stand is the scientific equivalent of writing 'Here be dragons' on an incomplete map. We did, however, have positive proof of the existence of antimatter particles here on Earth when scientists at the CERN particle physics laboratory in Switzerland created nine atoms of antihydrogen. The particles existed for just 40 nanoseconds.

FACT

The Universe is filled with cosmic background radiation, which is an echo of the Big Bang explosion. This can still be heard as white noise when you tune your radio between different stations.

The forces of nature

The behaviour of all matter in the Universe is governed by the four fundamental forces. These forces, or interactions as they are sometimes called, are gravity, the electromagnetic force, the strong nuclear force and a force known as the weak interaction.

The fundamental forces are currently believed to be transmitted via tiny particles called gauge bosons. Each of the forces has its own gauge boson and these are called gravitons (for gravity), photons (for the electromagnetic force), gluons (the strong nuclear force) and W, Z particles (weak nuclear force).

The Grand Unified Theory

Since the early twentieth century scientists have been attempting to find a theory that will link the four fundamental forces and perhaps prove that they are all aspects of a single, primary force. The quest for this Holy Grail of science is called the search for the Grand Unified Theory. Despite the best efforts of some of the science community's finest minds – Albert Einstein included – this theory has yet to be proved.

ABOVE Albert Einstein in 1935 lecturing to the American Association for the Advancement of Science on the theory that the universe might be infinite.

Gravity

Of the four fundamental forces, gravity is the one with which most people are familiar. It binds galaxies together and causes apples to fall on the heads of geniuses and ignoramuses alike. Gravity is essentially a consequence of mass. Any object possessing mass can exert a gravitational influence on any other object possessing mass. The larger the masses of the two objects, and the closer they are to each other, the greater the gravitational attraction between them will be.

Gravitons, which are believed to be the carrier particles for this force, have yet to be discovered. An alternative but not opposing view, which is expressed in Einstein's General Theory of Relativity, states that gravity is a consequence of the distortion of the fabric of space-time by objects possessing mass.

FACT
Gravity appears to be the weakest of the fundamental forces. Some mathematicians believe that this may be because it is a force that is felt across 11-dimensional space and as such is spread a bit thin in the four-dimensional version of the Universe that we can perceive.

The electromagnetic force

All particles that have an electric charge, such as the electron, are under the influence of the electromagnetic force. This force acts between the atoms and molecules of solid objects and gives them their rigidity. It also holds together the atoms in water molecules and, should these freeze, it holds the water molecules together to form ice.

Magnets also behave the way they do because of the electromagnetic force. The carrier particle for the electromagnetic force is the photon. These particles are the same ones from which visible light is 'made'.

FACT
The Irish writer James Joyce was the first person to use the word 'quark', in his novel *Finnegans Wake*.

The strong nuclear force

The strong nuclear force is at work in the nucleus of an atom, holding together the neutrons and protons. Protons have a positive charge and are constantly trying to repel each other – they would fly apart were it not for the presence of this force. In essence, the strong nuclear force is the glue that binds the nucleus of an atom together. Rearrangement of the particles that are held together by the strong nuclear force releases vast amounts of energy and it is in this way, through the rearrangement of hydrogen atoms into helium atoms, that the Sun and all other stars derive their energy.

The weak interaction

The weak interaction is probably the least interesting of all the fundamental forces. It doesn't bind galaxies together, it doesn't shine a light on anything or govern nuclear reactions in the core of a star, but it does glow in the dark.

It enables radioactive beta decay, which results in the break-up of the nucleus of an atom. By their very nature, radioactive atoms are unstable – their nuclei are home to too many neutrons. By virtue of the weak interaction, a neutron can change into a proton and in the process give off an electron (which in this instance is called a beta particle). What relevance does this have to your daily life? Well, it's the reason why your watch face glows in the dark.

RIGHT The rising Sun makes a dramatic backdrop for the space shuttle *Discovery's* rollout in 1997.

The search for dark matter

Astronomers have a very good idea of just how much mass there should be in the Universe. Unfortunately, the Universe doesn't seem to agree with them. In truth, based on the astronomers' calculations, over 90 per cent of the Universe appears to be missing – the astronomers are certain that the matter is out there, they just can't seem to lay their hands on it at the moment.

Back in the 1930s, when astronomers were measuring the speed of rotating galaxies, they were shocked to discover that most were spinning at more than twice the speed they'd expected. So fast, in fact, that they should have been flying apart. But they weren't. Could it be that some kind of invisible mass, some 'dark matter', was holding them together?

This dark matter, which is thought to be made of particles so small that they could pass easily through the human body without being detected, is now believed to be bound up in vast clouds of hot gas which surround clusters of galaxies in deep space.

Obviously, although this gas can be seen, the dark matter it masks cannot. But the fact that these gas clouds are caught in the gravitational influence of the galaxy clusters indicates that there must be a great deal more mass present than appears to be the case. If not, the gas would simply drift off into space. The latest calculations indicate that there must be at least four times as much invisible mass – or dark matter – in the clouds as there is visible matter.

At present astronomers are attempting to detect dark matter particles by setting up incredibly sensitive equipment at the bottom of deep mineshafts, well away from the distorting effects of cosmic radiation. Should dark matter ever be found it will provide conclusive proof of Einstein's general theory of relativity.

LOOKING AT
GALAXIES

What is a galaxy?

A galaxy is essentially an immense collection of stars which are held together by gravity. In among the stars one will also find lots of billowing gas clouds and dust as well as assorted planets, moons, asteroids, meteors, comets and perhaps even the occasional black hole. Galaxies range in size from just a few hundred light years across to enormous celestial beasts spanning over three million light years of space and stuffed with more than a trillion different stars. Scientists estimate that there are about 100 billion galaxies in the Universe. Rather than being spread out evenly, however, they tend to be grouped into superclusters.

Our knowledge of galaxies remains incomplete as it was only relatively recently that they were recognised as bodies which were composed of stars. Using greatly improved telescopes that were invented and produced during the early twentieth century, astronomers came to realise that what had once been thought of as fairly nearby nebulae were in fact far distant galaxies. This led to some previously catalogued deep-space objects being reclassified, with the Andromeda Nebula, for example, being renamed the Andromeda Galaxy.

Back in the eighteenth century the French astronomer Charles Messier created a system of classification that still has an influence on astronomy today; it is the reason why many galaxies still appear with the prefix 'M' (for Messier). A later listing, also in use now, was the New General Catalogue. This added the prefix NGC to all known galaxies.

In 1926, the American astronomer Edwin Hubble, who did so much to improve our understanding of the sheer scale of the Universe, categorised all the known galaxies according to shape. This approach imposed four basic models: irregular; elliptical; spiral; and barred spiral. Later, in the early 1960s, a new and more distant type of galaxy was discovered. Situated billions of light years from Earth, the quasar remains something of a mystery.

The Milky Way

Our Solar System is to be found in one arm of a spiral galaxy called the Milky Way. The nucleus of the galaxy is about 12,000 light years in diameter, but the disc in which it sits is eight times wider at 100,000 light years in diameter. This disc is 1,000 light years thick.

The Sun shares this galaxy with around 500 billion other stars, many of which are concentrated in a dense spherical nucleus near the centre of the spiral. The galactic centre is believed to be home to a fearsomely massive black hole called Sagittarius A★. The rest of the stars in the galaxy are scattered throughout the four 'arms' of the spiral, which are named Sagittarius, Orion, Perseus and Carina. The Sun (and, of course, the Earth) is located in the Orion arm, about 30,000 light years from the centre of the galaxy.

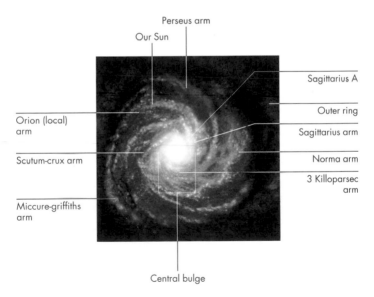

Perseus arm

Our Sun

Sagittarius A

Outer ring

Orion (local) arm

Sagittarius arm

Scutum-crux arm

Norma arm

3 Killoparsec arm

Miccure-griffiths arm

Central bulge

ABOVE Milky Way Galaxy, viewed from above.

BELOW Milky Way Galaxy.

The Sun orbits the centre of the Milky Way Galaxy at a speed of around 220 km (138 miles) per second but even at this speed it takes approximately 250 million years to complete a single orbit. It is thought that the Sun has orbited the centre of the galaxy somewhere between 15 and 23 times so far.

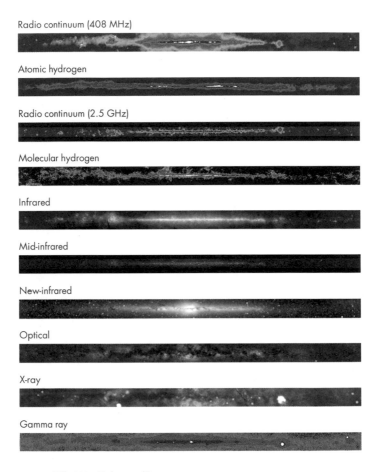

Radio continuum (408 MHz)

Atomic hydrogen

Radio continuum (2.5 GHz)

Molecular hydrogen

Infrared

Mid-infrared

New-infrared

Optical

X-ray

Gamma ray

ABOVE Milky Way Galaxy multiwave.

Our intergalactic neighbours

Our galaxy, enormous as it is, is only one of several in a relatively small cluster of about 22 galaxies that go to make up the Local Group. The Local Group occupies a region of space that is about three million light years across. Although this cluster has no real centre, the two most massive galaxies, ours and the Andromeda Galaxy, form the centres of two sub groups.

The Large and Small Magellanic Clouds are the nearest galaxies to our own. The Large Magellanic Cloud is 179,000 light years away while the Small Magellanic Cloud lies 210,000 light years away. The use of the word 'cloud' in their names is a throwback to the days when these galaxies were believed simply to be clouds of gas and dust drifting in space, a misconception that continued until improvements in telescope design revealed the truth.

After Andromeda and the Milky Way, the Large Magellanic Cloud and the magnificent spiral galaxy M33 are the next largest in size. The remaining galaxies are relatively small by comparison and are largely faint elliptical or irregular-shaped galaxies.

The Local Group is part of an even larger body called the Local Supercluster. This contains several hundred galaxy clusters and is arranged in a shape that can only be likened to that of a loose handful of string. The supercluster strings constitute the boundaries to voids that are truly incomprehensible in size. These voids separate superclusters from each other.

FACT
Why are there no external photographs of the Milky Way Galaxy in a book dedicated to astronomy? The reason is that no man-made object, let alone a camera, has ever travelled far enough out into the reaches of space to be able to take such a shot.

The Andromeda Galaxy

Also known as NGC 224 and M31, the Andromeda Galaxy is a spiral galaxy that shares many characteristics with our own Milky Way Galaxy. Originally thought to be a component of the Milky Way, Andromeda was first mentioned as early as 964 CE by the Islamic scholar Abd Al-Rahman Al Sufi in his *Book of Fixed Stars*. It was rediscovered in 1612 by the German astronomer Simon Marius, who found it using the newly invented telescope. Originally thought to be a cloud of gas and dust, it was not until the 1920s that the American astronomer Edwin Hubble showed that the Andromeda Galaxy is a separate galaxy in its own right lying beyond the boundaries of the Milky Way.

The Magellanic Clouds

Named after the Portuguese explorer Ferdinand Magellan, the Magellanic Clouds formed several billion years after the Milky Way Galaxy. Condensing from matter left over after the Big Bang, they contain many relatively young stars. This means that they provide an excellent opportunity for astronomers to study the formation and evolution of stars.

Back in 1987, the light from a supernova (exploding star) that had occurred 150,000 years ago finally reached us here on Earth, offering an unparalled opportunity to study the sudden death of a star that had previously been catalogued. This was the nearest supernova to Earth to occur in the last 200 years.

The Great Wall

Astronomers have identified a structure that is even larger than a supercluster. Called the Great Wall, it is made up of superclusters and a few scattered clusters. The Great Wall occupies a region of space measuring approximately 500 million light years by 300 million light years by 15 million light years. It is separated from the next 'wall' by an eerily empty region of space that spans some 400 million light years.

LEFT Large Magellanic Cloud, which is our nearest galactic neighbour.

Galaxies of the Local Group

Name of galaxy	Distance in light years	Name of galaxy	Distance in light years
Milky Way	0	Andromeda II	2,200,000
Large Magellanic Cloud	179,000	Andromeda III	2,200,000
Small Magellanic Cloud	210,000	M32	2,200,000
Draco	300,000	NGC 147	2,200,000
Carina	300,000	NGC 185	2,200,000
Sculptor	300,000	NGC 205	2,200,000
Sextans	300,000	M33	2,400,000
Ursa Minor	300,000	IC 1613	2,500,000
Fornax	500,000	DDO 210	3,000,000
Leo I	600,000	Pisces	3,000,000
Leo II	600,000	GR8	4,000,000
NGC 6822	1,800,000	IC 10	4,000,000
IC 5152	2,000,000	Sagittarius	4,000,000
WLM	2,000,000	Pegasus	5,000,000
Andromeda	2,200,000	Leo A	5,000,000
Andromeda I	2,200,000		

FACT

Situated at a distance of 2.2 million light years from Earth, the Andromeda Galaxy is the most distant object that is visible to the naked eye. The galaxy was observed by Persian astronomers as far back as 905 CE, and was described by Abd-al-Rahman Al-Sufi in his 964 CE *Book of Fixed Stars* – he referred to it by the utterly charming name of 'the little cloud'.

Types of galaxies

In the 1920s Edwin Hubble defined galaxies according to types based on their appearance. These 'types' were of four broad categories: irregular; elliptical; spiral; and barred spiral. Advances in telescope and camera technology, in particular the use of the charge-coupled device (CCD) which 'collects' light in a far more efficient way than the human eye could ever manage, means that we now have an unparalleled opportunity to view what are arguably the most magnificent structures in creation.

Irregular galaxies

Lacking any recognisable shape or structure, irregular galaxies tend to have less mass than other types of galaxies but are home to many bright young stars. Despite this, irregular galaxies make up less than five per cent of the brightest thousand or so galaxies and yet form about a quarter of all known galaxies.

ABOVE An irregular-shaped galaxy, the Small Magellanic Cloud.

Elliptical galaxies

As the name suggests, elliptical galaxies are shaped like ellipses. The ellipse may be anything from almost spherical in shape to something approaching the shape of a rugby ball. Unlike the other types of galaxies, the elliptical galaxies appear yellow in colour. This is because the process of star formation has generally all but stopped, which means that the stars in the galaxy are nearly all old red giant stars, tending for the most part to be more than 10 billion years old.

Barred spiral galaxies

In many respects similar to the spiral galaxies, barred spiral galaxies differ in one important detail: the bulging central nucleus is extended laterally in two directions to form a bar. This bar-shaped central nucleus can be a variety of sizes but the larger ones tend to have tightly wound arms while the smaller ones have arms that are loosely packed with stars. As the central bar rotates, the arms of the spiral appear to extend from each end of the bar. Although treated as a distinct type by Edwin Hubble, astronomers are increasingly beginning to recognise spiral and barred spiral galaxies as variations on the same theme.

ABOVE An example of a barred spiral galaxy showing a bar of stars, dust and gas.

Spiral galaxies

Truly beautiful to behold, spiral galaxies are centred on a bulging nucleus from which extends several trailing 'arms' of stars, from which they get their name. Older stars can be found in the dense nucleus with younger stars forming in the glowing pink nebulae of the far less dense arms. Vast amounts of gas and dust can also be found in the arms, some of which will ultimately condense to form new stars. The very oldest stars inhabit a sparsely populated 'halo' which envelops the nucleus.

ABOVE The M100 galaxy is a large spiral galaxy similar to our own Milky Way, containing over 100 billion stars. It is over 150 million light years away. The picture was taken in 1993 with the Wide Field and Planetary Camera 2 on board the Hubble Space Telescope.

When galaxies collide

It's tempting to think of galaxies as islands in space, fixed to a single spot around which they rotate. In reality, all the galaxies are flying through the Universe at tremendous speeds, with the galaxies furthest away from our own travelling fastest of all. Despite the vast distances separating galaxies – our nearest galactic neighbour is at least 170,000 light years away – galaxies occasionally collide . . . with results that go far beyond what might be described as catastrophic.

As two galaxies move closer together they become caught in each other's gravitational fields. Before long, visible streams of matter, initially nebulous gas and dust, can be seen pouring from one galaxy to the next. As this continues it has a dramatic effect on the appearance of the galaxies. Depending on the direction of drift the two galaxies may eventually merge to form a single, much more massive galaxy.

This is currently happening to galaxies NGC 4038 and NGC 4039, which together are known as the Antenna. Eventually, and inevitably, these two galaxies will combine to form a single galaxy.

Seyfert galaxies

Despite Edwin Hubble's best efforts, not all galaxies can be slotted into the four broad categories he defined. Since Hubble made his observations two other varieties of galaxy, each even more mysterious and exotic than anything known of in Hubble's day, have been discovered.

The first was recognised in 1943 by the American astronomer Carl K Seyfert. When observing the spectral emissions for a number of apparently ordinary spiral galaxies, Seyfert noticed a concentration of hot gases, especially hydrogen, at their central cores. This gas was expanding at tremendous speed – up to several thousands of kilometres per second.

Despite appearing to be quite normal in conventional photographs, these galaxies were clearly very strong sources of infrared radiation. Many were also powerful sources of radio waves and X-rays, suggesting the presence of a black hole in the nucleus of the galaxy. Since Seyfert made his discovery astronomers have noticed that about one per cent of all spiral galaxies are Seyfert galaxies. But it may yet prove to be the case that all spiral galaxies display Seyfert properties one per cent of the time.

ABOVE The Seyfert spiral galaxy NGC 7742.

FACT

Not all stars are held in galaxies; some are roaming free throughout space. Using the Hubble Space Telescope, astronomers discovered over 600 stars adrift between the galaxies of the Virgo cluster.

FACT

Despite the remarkable amount of energy that is released by these intense little galactic entities, no quasar is visible to the naked eye – hardly surprising given their distance from Earth. But in addition to radio waves and light waves, quasars are also powerful sources of X-rays, gamma rays, ultraviolet light and infrared radiation.

Quasars

While Seyfert galaxies might be thought of as a specific type of spiral galaxy, 1962 saw the discovery of a whole new – and truly ancient – type of galactic entity. Thought to be related to Seyfert galaxies, quasars are among the rarest and brightest objects in the Universe.

Classed as QSOs, which stands for 'quasi-stellar objects' (the name 'quasar' is a shortening of 'quasi-stellar radio source'), these tiny galaxies are never more than a couple of light years across yet shine a thousand times brighter than galaxies with a diameter of over 100,000 light years.

The incomparable brilliance of the quasar means that they can be observed from Earth despite being up to 10 billion light years away. The enormous amounts of radiation required for such a feat are released from just a small area at the centre of the quasar. Many astronomers believe that this energy is being released by gases accelerating to speeds approaching those of light as they are sucked into a super massive black hole.

The brightest quasar so far discovered is also the nearest one to us. Situated two billion light years from Earth, NGC 3C 273 has a nucleus that is currently expanding at very near the speed of light, a fact which points to the presence of a black hole at the centre of this quasar. Quasars are so bright that some can be observed even though the energy coming from them began its journey to us five billion years before our Sun was born.

SPACE
EXPLORATION

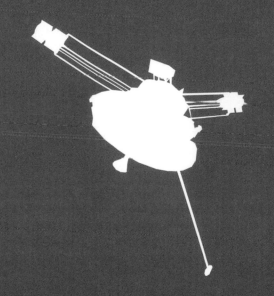

How rockets work

A rocket engine is really just a firework built on a grand scale. It requires fuel, which might be liquid hydrogen, and a source of oxygen, the gas needed for combustion. This mixture of fuel and oxidizer is known as the propellant and it is burned to produce hot gases which expand rapidly. This rapidly expanding gas is forced through a nozzle at the base of the rocket in order to propel it in an upwards direction.

This is not a new idea. The Chinese were making rockets – for military and display purposes – as early as the eleventh century. The principle on which the rocket lifts off the ground and continues its journey skywards was described by the English scientist Isaac Newton in the seventeenth century. His Third Law of Motion states that for every action (the downward thrust of the gases) there is an equal and opposite reaction (lift-off).

Once a rocket has left the ground it relies on the thrust from the gas propellant to carry on climbing upwards. The rate of climb depends on the amount of thrust the rocket engines produce and the mass of the rocket itself – the more the rocket weighs the more thrust it needs to travel the 80 km (50 miles) or so into space. The rocket will continue to accelerate for just as long as the rocket engines are burning.

Types of rocket engine

Rocket engines are of two main types: solid fuel or liquid propellant. Some, such as the space shuttle, may be equipped with both for take-off. As an alternative, extremely efficient nuclear-powered rockets have been proposed. But so far fears over the consequences of an accident occurring with such an engine have kept this scheme at the drawing-board stage.

RIGHT Space shuttle *Atlantis* leaps from the steam and smoke billowing across Launch Pad 39B after an on-time lift-off of 3:46 p.m. EDT on mission STS-112.

Early rocketry

As far back as 1903 the Russian scientist Konstantin Tsiolkovsky suggested that rockets which use liquid propellants could reach space. This may not seem like an especially significant suggestion but it was the first seriously proposed change to the design of rockets to have occurred in nearly 900 years.

After much work and many failures, a successful design for a liquid fuel rocket was demonstrated in 1926 by the American scientist Robert Goddard. But it was the onset of war, however, that provided the impetus – and

ABOVE Robert Goddard with an early rocket design which he hoped would provide greater stability in flight.

an enormous injection of funding – required to build rockets that had the power to reach into outer space.

Von Braun and the *V2*

In 1944 the German military launched a terrible new weapon on the cities of Paris, Antwerp and London. Travelling faster than the speed of sound – and so arriving before anyone could be warned of its approach – the *V2* rocket was able to carry a 1,000 kg (2,200 lb) warhead over a distance of 320 km (200 miles) and, most significantly, at a height of up to 160 km (100 miles). Germany may have failed in its bid to conquer the world, but as part of its war effort it had conquered space. The man who led the team that developed the *V2* rocket at Peenemünde in Germany was Werner von Braun. At the end of the war he surrendered to American forces and went on to develop the *Saturn V* rocket, an essential part of the *Apollo* program that culminated in the successful Moon landing of 1969.

Korolyov and the Soviets

The most important figure in the early decades of the space age was a little-known Ukrainian-born engineer called Sergei Korolyov. He was responsible for the development of all Soviet spacecraft up to and including the *Soyuz* series and played a vital role in the launch of *Sputnik 1*, the first satellite in space. He also guided the first human spaceflight (by Yuri Gagarin in 1961) and the first spacewalk (by Aleksei Leonov in 1965).

BELOW Evolution of Soviet rockets.

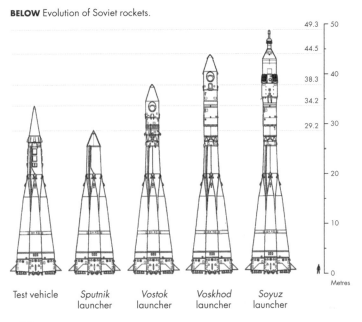

| Test vehicle | *Sputnik* launcher | *Vostok* launcher | *Voskhod* launcher | *Soyuz* launcher |

FACT

The first satellite in space was called *Sputnik*. The word is Russian for 'travelling companion', or 'satellite'.

Timeline of rocketry and space exploration

Date	Event
c.1100	China develops the first rockets, essentially large fireworks which are used for military and display purposes.
c.1700	Rocket technology rediscovered in the West; rockets appear on European battlefields.
c.1792	Indian prince Tipu Sultan uses a new and more powerful rocket against the British at Seringapatam.
c.1810	Artillery officer William Congreve improves on the design of the rockets he saw first-hand at Seringapatam, building one that will travel over two thousand metres to its target. His rockets are used to attack Boulogne, Copenhagen and Danzig during the Napoleonic Wars and in the British attack on Fort McHenry, near Baltimore, in 1814.
c.1850	British engineer William Hale eliminates the dead-weight of the flight-stabilising guide stick (a feature of previous rocket designs) by adding jet vents at an angle which make the rocket spin, thus stabilising its flight through the air.
1861–65	Rockets used with varying degrees of success during the American Civil War.
c.1900	Swedish engineer Wilhelm Unge invents an 'aerial torpedo', which is designed to be used as a surface-to-air weapon against dirigibles.
1915	Konstantin Tsiolkovski designs, but does not build, the first liquid-fuel rocket.
1920s	Amateur rocket-enthusiast clubs open up all over Europe and the United States.
1930s	The military assume control of the main rocket clubs in Germany.
1944	The German military launch the *V2* rocket weapon against Paris, Antwerp and London.
c.1945	The USSR and the USA begin the race to build rockets capable of carrying nuclear weapons.
Oct 1957	The USSR successfully launches *Sputnik 1*, the first artificial satellite in space.
Nov 1957	The USSR's *Sputnik 2* mission puts a dog, Laika, into space. The animal dies shortly after take-off from overheating and stress.

Date	Event
Jan 1959	USSR's *Luna 1* mission escapes Earth's gravity.
Sep 1959	*Luna 2* reaches the Moon.
July 1959	*Luna 3* photographs the far side of the Moon.
May 1960	*ECHO 1*, the first communications satellite, is launched by the United States.
31 Jan, 1961	A chimpanzee named 'Ham' is sent into space and returned alive as part of the US *Mercury* program.
12 April, 1961	Yuri Gagarin becomes the first man in space.
5 May, 1961	Alan Shepard becomes the first American in space.
20 Feb, 1962	John Glenn becomes the first American to orbit the Earth.
16 Jun, 1963	Valentina Tereshkova becomes the first woman in space.
18 Mar, 1965	Soviet cosmonaut Alexei Leonov makes the first spacewalk.
1966	The USSR lands a spacecraft on the surface of Venus.
21–27 Dec, 1968	The crew of *Apollo 8* orbit the Moon.
20 July, 1969	*Apollo 11* astronauts Neil Armstrong and Buzz Aldrin become the first men to set foot on the Moon. Fellow crew member, Michael Collins, remains in orbit around the Moon.
19 April, 1971	The Soviets launch the first space station, *Salyut 1*.
27 Nov, 1971	The Soviets crash-land a probe on Mars.
1981	The United States launches the space shuttle.
1990	The Hubble Space Telescope is launched.
2000	The International Space Station opens for business.

FACT

The British attack on Fort McHenry, near Baltimore, in 1814 was carried out using Congreve's rockets. The rockets inspired Francis Scott Key to write 'the rockets red glare, the bombs bursting in air' for the US national anthem.

BELOW Map of the world's launch sites.

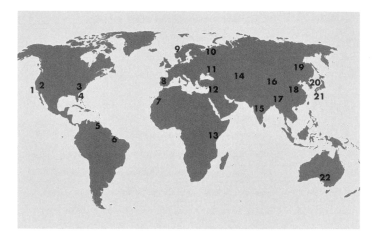

1	Vandenberg, USA	12	Palmachim, Israel
2	Edwards, USA	13	San Marco, Kenya
3	Wallops Island, USA	14	Baikonur, Russia
4	Cape Canaveral, USA	15	Sriharikota, India
5	Kourou, French Guiana	16	Jiuquan, China
6	Alcantara, Brazil	17	Xichang, China
7	Hammaguir, Algeria	18	Taiyuan, China
8	Torrejon, Spain	19	Svobodny, Russia
9	Andoya, Norway	20	Kagoshima, Japan
10	Plesetsk, Russia	21	Tanegashima, Japan
11	Kapustin Yar, Russia	22	Woomera, Australia

Launch sites are few and far between with some open to the public, while others remain top-secret. The locations for spaceports are decided according to both political and technical realities. The most used ports are Cape Canaveral, Vandenberg, Baikonur, Plesetsk, Kourou, Tanegashima, Jiuquan, Xichang and Sriharikota.

Competition for the world's largest space programme remains intense, with the United States and Russia heading the race, followed by the ESA (European Space Agency), France, Japan, Germany, Italy, India, the United Kingdom, Canada, Brazil, Belgium and Spain.

The space race

World War II saw the rise of two superpowers, the USSR and the United States. In the aftermath of war, each competed with the other to build evermore sophisticated rockets capable of carrying nuclear weapons to the enemy's territory. Initially, and by virtue of superior design, the USSR led the way. In 1957, they succeeded in placing the first artificial satellite in space and topped this remarkable feat in 1961 by launching cosmonaut Yuri Gagarin on a trajectory that took him, briefly, into space.

By now, the United States was losing face and everyone knew it. Fearing the potential of unstoppable space-based weapons systems, the United States began to pour money into research and development. Faced with the need for a face-saving exercise in the wake of the disastrous Bay of Pigs incident of 1961, US President John F Kennedy pledged to place a man on the Moon before the end of the decade. On 20 July, 1969 Neil Armstrong left the very first human footprint on the Moon and uttered the immortal line, 'That's one small step for a man, one giant leap for mankind.' At the final hurdle, the USSR had lost the space race.

ABOVE JFK dedicating his nation to the race to space.

FACT
Because there is no atmosphere on the Moon, Neil Armstrong's first footsteps will not blow away. They are preserved in the dust that covers the Moon's surface and will remain there until someone – or something – in the future erases them.

ABOVE Yuri Gagarin on 12 April, 1961 on the *Vostok 1* flight.

Vostok

Between April 1961 and 1963 the USSR launched six manned missions to Earth orbit as part of the *Vostok* program. The first of these missions saw Yuri Gagarin become the first human in space. *Vostok 2* made 17.5 orbits of the Earth. *Vostok 3* and *Vostok 4* were launched together and orbited within sight of each other. *Vostok 5* was launched on 14 June, 1963 followed two days later by *Vostok 6*, which carried the first woman cosmonaut, Valentina Tereshkova, into space.

Mercury

Mercury, the US answer to the *Vostok* program, began with its first official launch, *Mercury 3*, on 5 May, 1961 taking Alan Shephard on a suborbital flight lasting 15 minutes and 22 seconds. The flight of *Mercury 4*, carrying Virgil 'Gus' Grissom and launched the same year, lasted an extra 15 seconds. *Mercury 5* launched in November 1961 and was an orbital flight crewed by a chimpanzee. *Mercury 6* followed in February 1962 with John Glenn at the controls. *Mercury 7* and *Mercury 8* followed in May and October of that year with the final flight of the program launching in May 1963.

ABOVE *Project Mercury astronauts, 13 September, 1968.*

Project Gemini

Between 1963 and 1966 the United States launched 10 two-man missions into Earth orbit in a 12-stage project known as *Gemini*. The project was designed to test the ability of astronauts to manoeuvre themselves and their equipment in the weightless environment of space. It also helped NASA to develop the techniques needed for orbital rendezvous and docking, procedures that were to play a vital role in the subsequent *Apollo* mission to the Moon.

Gemini 1 & 2. *In 1964 and 1965, NASA launched two preliminary spacecrafts to test that all the technology functioned properly. Both launches were successful.*

Gemini 3. *Launching on 23 March, 1965, this was the first US spacecraft to carry two men into space.*

Gemini 4. *This was launched on a four-day mission on 3 June, 1965. It was during this flight that US astronaut Ed White became the first American to walk in space: he left the **Gemini** capsule for 22 minutes.*

Gemini 5. *Its crew set a new record when it orbited the Earth for nearly eight days starting on 21 August, 1965.*

Gemini 6. *The launch vehicle exploded on take-off but finally achieved orbit on 15 December, 1965.*

Gemini 7. *This reached space on 4 December, 1965.*

Gemini 8. *Piloted by Neil Armstrong, **Gemini 8** achieved the first successful space docking after it was launched on 16 March, 1966.*

Gemini 9. *On 3 June, 1966 **Gemini 9** reached space where astronaut Eugene 'Gene' Cernan, later the last man to walk on the Moon, took his first walk in space.*

Gemini 10. *This launched without incident on 18 July, 1966.*

Gemini 11. *Launched on 12 September, 1966.*

Gemini 12. *The last mission of the **Gemini** program was launched on 11 November, 1966. The way to the Moon was now clear and would be achieved via a new set of missions: the **Apollo** project.*

LEFT *Gemini-Titan 4* launch. The flight included the first space walk by a US astronaut on 6 March, 1965.

Past Space Missions

Date	Name (nationality)	Mission objective	Mission outcome
4 October, 1957	*Sputnik 1* (USSR)	First Earth orbiter.	Successfully completed.
3 November, 1957	*Sputnik 2* (USSR)	First Earthling in space: Laika the dog!	Died after a week in space.
12 April, 1961	*Vostok 1* (USSR)	First man in space. Yuri Gagarin orbited Earth once.	Successfully completed.
16 June, 1963	*Vostok 6* (USSR)	First woman in space: Valentina Tereshkova.	Successfully completed.
18 March, 1965	*Voskhod 2* (USSR)	First spacewalk: Alexei Leonov.	Successfully completed.
17 August, 1970	*Venera 7* (USSR)	First Venus lander.	Successfully completed.
14 May, 1973	*Skylab* (USA)	Space station (carried the first lavatory and first shower in space).	After successfully completing mission brief, burned up in atmosphere.
20 August, 1975	*Viking 1* (USA)	Martian lander.	Successfully completed.
12 April, 1981	*STS-1 Space Shuttle Columbia* (USA)	First shuttle mission: systems check.	Successfully completed.
2 July, 1985	*Giotto* (Europe)	Visited Halley's Comet.	Successfully completed.
28 January, 1986	*STS-51-L Space Shuttle Challenger* (USA)	Deployment of tracking satellite and Halley's Comet observer.	Exploded on launch, all astronauts lost.
25 September, 1992	*Mars Observer* (USA)	Mars orbiter.	Contact lost.
4 December, 1996	*Mars Pathfinder* (USA)	Mars lander and rover.	Successful.
3 January, 1999	*Mars Polar Lander* (USA)	Mars lander.	Mission failure.
16 January, 2003	*STS-107 Space Shuttle Columbia* (USA)	16-day mission dedicated to research in physical, life and space sciences.	Exploded on re-entry, all astronauts lost.

The *Apollo* program

27 January, 1967 should have marked the day that the *Apollo* program finally left the launch pad. As *Apollo 1* sat on the launch pad its crew, Virgil 'Gus' Grissom, Edward White and Roger Chafee, carried out last-minute checks of the craft's command systems. By now the three men, who had been sealed into the cramped capsule, were breathing an atmosphere of pure oxygen and praying that nothing would go wrong.

A single spark triggered an electrical fire. In the pressurised pure-oxygen atmosphere of the capsule there was no way of stopping the blaze. In seconds all three men were engulfed in the blowtorch blaze with no hope of escape or rescue. By the time they were freed from the seats in which they had died, their bodies had been burned beyond recognition, victims of bad design and even worse luck. The next *Apollo* capsule would have an escape hatch.

The stated aim of the *Apollo* program was to place a man on the Moon before the end of the 1960s. At the time the announcement was made only two people had ever travelled into space, and even then for only a few minutes in total. But by the time that the crew of *Apollo 1* approached their vehicle for take-off a good deal of groundwork had been carried out, with first one-man (*Mercury* program) and then two-man (*Gemini* program) vehicles taking to space in what were essentially preparatory flights for the journey to the Moon.

After the *Apollo 1* disaster the next two flights were cancelled. Finally, test flights for the *Apollo* program began again on 9 November, 1967 with the launch of *Apollo 4*.

The *Apollo* Missions

Apollo 1

Launch date: 27 January, 1967

Astronauts Virgil Grissom, Edward White and Roger Chafee all died tragically when an electrical fire broke out on the spacecraft just before take-off.

Apollo 4
Launch date: 9 November, 1967

The *Apollo* program resumed with an unmanned mission which saw NASA test the first *Apollo-Saturn V* rocket combination in flight.

Apollo 5
Launch date: 22 January, 1968

This unmanned mission saw the first test flight, in Earth orbit, of the Lunar Module (the part of the craft that would eventually land on the Moon).

Apollo 6
Launch date: 4 April, 1968

The final unmanned mission saw further testing of the *Saturn V* rocket that would eventually send the *Apollo* spacecraft to the Moon.

Apollo 7
Launch date: 11 October, 1968

Astronauts Walter Schirra, Walter Cunningham and Donn Eisele tested and evaluated *Apollo's* navigation and control systems over the course of 11 days while in Earth orbit.

Apollo 8
Launch date: 21 December, 1968

The crew of *Apollo 8* – Frank Borman, James Lovell and William Anders – became the first people in history to escape Earth's gravity. They went on to fly around the Moon, orbiting it 10 times on Christmas Eve 1968 while observing possible landing sites for later *Apollo* missions.

Apollo 9
Launch date: 3 March, 1969

Astronauts James McDivitt, David Scott and Russell Schwieckart orbited the Earth 152 times while testing out docking procedures and the *Apollo* Life Support System, a specially designed spacesuit.

Apollo 10

Launch date: 18 May, 1969

Essentially a full dress rehearsal for the actual Moon landing, astronaut Thomas Stafford remained in orbit around the Moon while John Young and Eugene Cernan flew the *Lunar Module* to within 15 km (9 miles) of the surface of the Moon, all the while resisting the tremendous urge to land the craft and enter the history books as the first men on the Moon.

Apollo 11

Launch date: 16 July, 1969

The big one! On 20 July, 1969 an estimated one billion people were glued to their TV sets, witnesses to the remarkable sight of astronaut Neil Armstrong taking his first steps on the Moon and uttering one of the most famous (and famously misquoted) lines in history: 'That's one small step for a man; one giant leap for mankind.' He was followed on to the surface of the Moon some 20 or so minutes later by 'Buzz' Aldrin, who never really got over the disappointment of being the second man on the Moon. The other astronaut, Michael Collins, remained in orbit while history unfolded beneath him.

Apollo 12

Launch date: 14 November, 1969

Astronauts Charles Conrad and Alan Bean spent nearly 32 hours on the Moon where they collected samples and carried out experiments before blasting off to rejoin Richard Gordon in orbit around the Moon in the command module. They arrived back at Earth on 24 November.

ABOVE *Apollo 11* boot print for 'Buzz' Aldrin on 20 July, 1969.

FACT

Although an estimated one billion people watched the Moon landing, by the end of the *Apollo 12* mission viewers all over Europe and America were phoning in to complain that their regular soap operas were being disrupted by coverage of events on the Moon.

Apollo 13

Launch date: 11 April, 1970

At about 1:00 p.m. on 13 April, *Apollo 13* suffered disaster when an oxygen tank exploded in the service module. More than halfway to the Moon, the crew of James Lovell, John Swigert and Fred Haise found themselves trapped in the icy cold depths of space with no possibility of rescue aboard a ship that could no longer supply them with power, oxygen or water. Surviving on the Lunar Module's limited supply of oxygen, the men managed to return to Earth relatively unharmed despite the appalling conditions on board. They splashed down in the Pacific Ocean on 17 April, 1970.

ABOVE *Apollo 13* view of damaged service module.

Apollo 14
Launch date: 31 January, 1971

Astronauts Alan Shepard and Edgar Mitchell landed safely on the Moon and carried out many of the experiments that had originally been part of the *Apollo 13* mission. Stuart Roosa remained in orbit around the Moon.

Apollo 15
Launch date: 26 July, 1971

As part of the first mission that used the *Lunar Roving Vehicle*, astronauts David Scott and James Irwin drove across nearly 28 km (17.5 miles) of the Moon's surface while conducting detailed studies of the area. Meanwhile, astronaut Alfred Worden remained in orbit around the Moon.

Apollo 16
Launch date: 16 April, 1972

Astronauts John Young and Charles Duke took core samples and seismic soundings during the 20 hours they spent outside the *Lunar Module*. Meanwhile, astronaut Thomas Mattingly remained in orbit around the Moon so that he could work on mapping the pitted surface.

Apollo 17
Launch date: 7 December, 1972

Astronauts Harrison 'Jack' Schmitt and Eugene Cernan were the last men to walk on the Moon as part of a sample-gathering mission which saw them travel over 30 km (19 miles) across the surface of the Moon in the *Lunar Rover*. Astronaut Ronald Evans remained in orbit around the Moon.

Space stations

Placed in high orbit around the Earth, space stations are essentially little more than science labs with spectacular views. Some provision is made for the crew, but distinctly Spartan accommodation aside,

FACT
The very first space shuttle was named *Enterprise* in honour
of the spaceship in the television series *Star Trek*. Although
it never made it into orbit, it was being used for test
landings as far back as 1977.

these vehicles offer an unparalleled opportunity to conduct
experiments in a weightless environment that is also free from the
effects of the Earth's atmosphere.

Solar panels are an obvious feature of all space stations, generating
electricity to power essential systems while reducing the need for
heavyweight battery packs. Shielding protects the crew from the
worst effects of solar radiation and space debris and a docking port
allows supply craft to deliver cargo. This also means that the crew of
the space station can be replaced without the vessel having to return
to Earth.

In the future scientists have suggested placing space stations
outside of Earth orbit so that they can act as convenient rest
and re-supply stops on the journey to planets such a Mars and,
perhaps, beyond.

ABOVE *Salyut 2* series space station.

The space shuttle

Launched in 1981, the space shuttle was a revolutionary craft: it was the first-ever reusable spaceship. The design of the shuttle breaks down into three principal components, the orbiter space plane (the part most people think of when they imagine the shuttle), huge solid-fuel rocket boosters and an external fuel tank. Of these three components only the fuel tank is not reused after a flight.

Once in space, the shuttle is manoeuvred between orbits, the altitude of which can be up to 390 km (244 miles), using the orbital manoeuvring system at the rear of the craft. Smaller movements are effected using tiny jets.

A large robotic device called the remote manipulator arm is used to move the shuttle's payload, which might be a satellite, into space or to collect it for repair or recovery. Alternatively, the payload bay can be fitted with *Spacelab*, a special laboratory designed and built by the European Space Agency for use in Earth orbit.

ABOVE Looking down into the payload of the space shuttle in flight.

Satellites

For our purposes a satellite is an object that sits in orbit around Earth. It was the launch of a satellite, the tiny *Sputnik 1*, which triggered the space race. In July 1955, the White House declared its intention to launch a satellite into the Earth's orbit by the spring of 1958, only to be beaten to it by the Soviet Union, which launched *Sputnik 1* on 4 October, 1957. Over the next decade or so, the United States responded by investing trillions of dollars in order to regain supremacy in space, and while much of it was invested in high-profile manned missions to the Moon, more than a little was spent on the far less glamorous, but much more practical, satellite program.

Military satellites are designed for monitoring and communication purposes, such as GPS (Global Positioning System) signals and the guidance of 'Smart' missiles to their targets. The first commercial satellites were developed in order to aid communication, providing the opportunity to broadcast signals reliably beyond the horizon by bouncing them off an object fixed in orbit around Earth. Most communications satellites are held in high geostationary orbits where they enable everything from television broadcasts to long distance telephone conversations.

Types of orbits	Distance above Earth's surface
Low Earth Orbit (LEO)	200–1,200 km (124–746 miles)
Medium Earth Orbit (MEO)	1,200–35,286 km (746–21,927 miles)
Geosynchronous Orbit (GEO)	35,786 km (22,237 miles)
High Earth Orbit (HEO)	35,786+ km (22,237+ miles)

FACT

The largest artificial satellite in orbit around Earth is the International Space Station.

ABOVE *Explorer 1*, the first US satellite, successfully launched on 31 January, 1958.

FACT

The geostationary orbit (also known as the Clarke Orbit) is the brainchild of science fiction writer Arthur C Clarke. In 1945, he suggested that a satellite could be made to orbit the Earth at the same speed as Earth rotates on its axis, effectively fixing the satellite over one spot on the planet.

Probes

Probes, although often similar in size to satellites, are unmanned spacecraft designed to carry cameras and monitoring equipment beyond Earth's orbit to distant planets and moons. Once there, the probe can be controlled remotely, or simply follow a pre-programmed set of instructions, making whatever observations are necessary before broadcasting the information back in the form of radio waves to receiving stations here on Earth.

The earliest successful probe was the Soviet Union's *Luna 1*, which reached the surface of the Moon in 1959 – a full ten years before the United States landed a man on the surface.

Arguably, the most successful probe so far has been *Pioneer 10*, which was launched by NASA on 2 March, 1972. Originally designed to observe Jupiter, it became the first man-made object to leave the Solar System when, in 1983, it passed beyond the orbit of Pluto and into a region of space that still remains a mystery.

Scientists finally abandoned the probe after it became apparent that its last signal had been received on 22 January, 2003. At the time of broadcast, the probe was approximately 12 billion kilometres (7.6 billion miles) away and its signal, which was travelling at the speed of light, took eleven hours and twenty minutes to reach Earth.

Probes designed to explore our Solar System can be powered using solar panels. Those going far from the Sun, however, require electricity generated using the heat from radioactive materials.

Lunar probes	Function
Luna program	Soviet lunar exploration (1959–1976).
Ranger program	US lunar hard-landing probes (1961–1965).
Zond program	Soviet lunar exploration (1964–1970).
Surveyor program	US lunar soft-landing probe (1966–1968).
Lunar Orbiter program	US lunar orbital (1966–1967).
Lunokhod program	Soviet lunar Rover probes (1970–1973).
Muses-A mission (*Hiten* and *Hagoromo*)	Japanese lunar orbital and hard-landing probes (1990–1993).
Clementine	US lunar orbital (1998).
Lunar Prospector	US lunar orbital (1998–1999).
Smart 1	European lunar orbital (2003).
LUNAR-A	Japanese lunar orbiter and penetrators, launch scheduled for 2004 but delayed ever since.
SELENE	Japanese lunar orbiter and lander, launch postponed to 2007.

Mars probes	Function
Zond program	Failed Soviet flyby probe.
Mars probe program	Soviet orbiters and landers.
Viking program	Two US orbiters and landers (1974).
Phobos program	Failed Soviet orbiters and *Phobos* landers.
Mars Pathfinder	Lander and wheeled robot (1996).
Mars Surveyor '98 program	Failed US probes.
Mars Odyssey	US orbiter.
Mars Observer	Failed US Mars orbiter.
Mars Express (*Mars Express Orbiter* and *Beagle 2*)	European orbiter and failed lander (2003).
Mars Exploration Rovers	US rovers (2004–present).
Mars Reconnaissance Orbiter	USA (launched 2005).
Mars Science Laboratory	USA (to be launched 2009).

Current and Future Space Missions

Date	Mission name	Mission objective	Status
August 1977	Voyager 1 (USA)	Jupiter and Saturn flyby.	Successful and still in operation.
September 1977	Voyager 2 (USA)	Jupiter, Saturn, Uranus and Neptune flyby.	Successful and still in operation.
25 April, 1990	Hubble Telescope (USA)	First and flagship mission of NASA's Great Observatories program.	Initial flaw in mirror corrected by shuttle mission; Hubble still in operation.
7 November, 1996	Mars Global Surveyor (USA)	Mars orbiter.	Still in operation.
15 October, 1997	Cassini (USA)	Saturn orbiter.	Successful arrival July, 2004.
15 October, 1997	Huygens (ESA) (onboard Cassini)	Titan probe.	Landed on Titan 14 January, 2005.
3 July, 1998	Nozomi (Japan)	Orbital survey.	Mission failure. Currently in useless orbit around the Sun.
7 April, 2001	Mars Odyssey (USA)	Orbital survey.	Still in operation.
2 June, 2003	Mars Express (ESA)	Mars orbiter and lander (Beagle 2 – UK).	Successful arrival 25 Dec, 2003. Still in operation.
10 June, 2003	Spirit (Mars Exploration Rover A) (USA)	Mars Exploration Rover.	Currently traversing Mars.
8 July, 2003	Opportunity (Mars Exploration Rover B) (USA)	Mars Exploration Rover.	Currently traversing Mars.
27 September, 2003	Smart 1 (ESA)	Lunar probe.	Currently in lunar orbit.
26 February, 2004	Rosetta (ESA)	Comet rendezvous.	Not available.
10–30 August, 2005	Mars Reconnaissance Orbiter (USA)	Orbital survey.	Not available.

Date	Mission name	Mission objective	Status
11 January, 2005	*Venus Express* (ESA)	Orbital survey.	Not available.
October 2007	*Kepler* (USA) planet finder	Terrestrial.	Not available.
October–December 2007	*Phoenix* (USA) (water search)	Mars lander.	Not available.
October–December 2009	*Mars Science Laboratory* (USA)	Mars lander and rover.	Not available.
1 January, 2011	*BepiColombo* (ESA)	Mercury orbital survey.	Not available.

ABOVE *Viking 2* on Mars' Utopian plain, 3 September, 1976.

Index

Picture credits

The publishers would like to thank the following for permission to reproduce images.
 NASA: pp 6, 21, 22, 33, 34, 36, 39, 42, 44, 46, 48, 49, 50, 51, 52, 53, 54, 55, 56, 57, 58, 59, 60, 61, 65, 67, 69, 71, 82, 83, 84, 85, 86, 87, 88, 92, 97, 102, 104, 106, 107, 110, 111, 112, 113, 114, 115, 119, 121, 126, 127, 128, 129, 132, 133, 135, 136, 139, 140
 Science and Society Picture Library: pp 9, 10, 11, 30, 31, 94, 120
 Richard Burgess (illustrator): pp 17 (b), 20, 26, 29, 32, 38, 40, 43, 45, 74, 75, 76, 77, 80, 101

Bibliography

Astronomy: A Beginner's Guide to the Universe. *Prentice Hall*, 2003
Astronomy for Dummies. *Hungry Minds Inc.*, 2005
National Geographic Encyclopedia of Space. *National Geographic Books*, 2005
The Right Stuff. *Vintage*, 2005
The Physics of Star Trek. *Basic Books*, 1995
A Man on the Moon: the Voyages of the Apollo Astronauts. *Time Life*, 1999

Websites

NASA Space technology and the Universe in general delivered by the people who have been there, done it, planted flags and brought back some lovely pictures. *www.nasa.gov*

Hubble Space Telescope Home to all of the astonishing Hubble Space Telescope images and more, with regular updates as they occur. *www.hubblesite.org*

Space.com Up-to-date space news and general information. *www.space.com*

The Particle Physics and Astronomy Research Council A more advanced site for those who wish to delve deeper into the subject. *www.pparc.ac.uk*

Royal Observatory at Greenwich FAQs and educational activities. *www.nmm.ac.uk*

European Space Agency Space technology and the latest news about Europe's attempts to conquer space. *www.esa.int*

The Science of Aliens The London Science Museum's take on the science of the search for alien life, which is linked to an award-winning, world-touring exhibition and includes fun features, interactive games and downloadable activities. *www.sciencemuseum.org*